Okavango Delta
Floods of Life

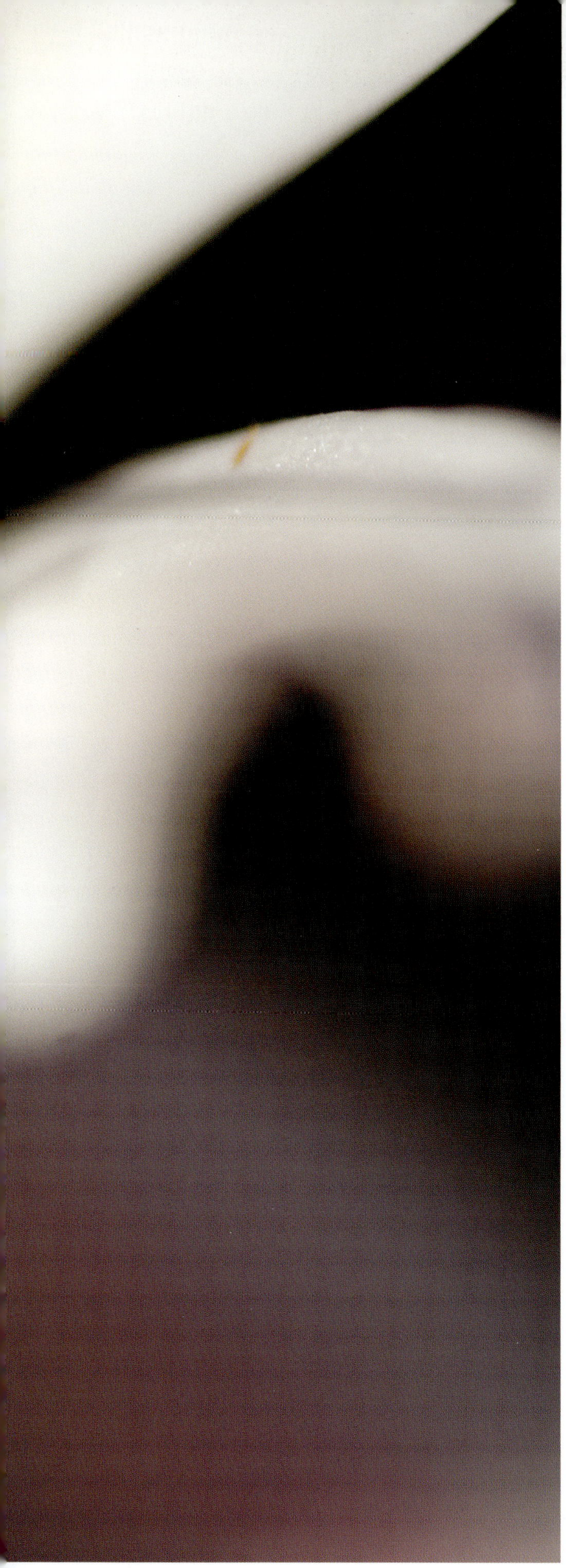

Okavango Delta: Floods of Life

John Mendelsohn, Cornelis vanderPost, Lars Ramberg, Mike Murray-Hudson, Piotr Wolski & Keta Mosepele

IUCN

International Union for the Conservation of Nature

European Union

THUTO KE THEBE

UNIVERSITY OF BOTSWANA
Harry Oppenheimer
Okavango Research Centre

This publication has been made possible with financial support from the European Union under grant Contract: EuropeAid/ENV/2004-81917, as part of an IUCN managed project, and with in-kind support from HOORC (Harry Oppenheimer Okavango Research Centre).

Published by: RAISON (Research & Information Services of Namibia) for IUCN and HOORC.

Project management: William Darwall (IUCN), Cornelis vanderPost (HOORC) and John Mendelsohn (RAISON)

First published: 2010

Citation: Mendelsohn JM, vanderPost C, Ramberg L, Murray-Hudson M, Wolski P & Mosepele K. 2010. *Okavango Delta: Floods of Life*. RAISON (Windhoek, Namibia).

ISBN: 978-99916-780-5-4

EAC ISBN: 9789991678054

Cover design by: Patricia Lynch-Blom

Layout by: Patricia Lynch-Blom

Printed by: Craft Print, Singapore

Cover image taken from NASA space shuttle mission STS043 on the 8th of August 1991

Contents

Acknowledgements

Many people gave the production of *Okavango Delta: Floods of Life* their support. Special credit is due to those who provided particular contributions of information, time and material support: Patricia Blom, Barbara Curtis, Helen Dallas, William Darwall, Bonnie Galloway, Pete Hancock, Mike Holding, Jessica Jones, Lucy Kemp, Jens Kipping, Naidu Kurugundla, Markus Lindholm, Anson Mackay, Spike McCarthy, Celia Mendelsohn, Nkobi Moleele, Belda Mosepele, Mark Muller and Ali Flatt, Richard Randall and Sue Ringrose.

Other important assistance was given by Rene and Marije Allemekinders, Jon Barnes, Brad Besterlink and Andy Crawford, Hans Beuster, Caspar Bonyongo, Lin Cassidy, Masego Dhliwayo, Louis du Preez, Birgitta Farrington, Richard Fynn, Clinton Hay, Don Kgathi, Martie MacFarlane, Lapo and Innocent Magole, Coleen Mannheimer, Wellington Masamba, Joseph Mbaiwa, Tico McNutt, Martin and Stephie Mendelsohn, Lesego Mmualefe, Monica Morrison, Sekgowa Motsumi, Chaminda Rajapakse, Vince Schacks, Roger Swart, Eliot Taylor and Ben van der Waal.

Staff at the HOORC library made special efforts to track down material that was otherwise hard to find. The assistance of the Central Statistics Office, Department of Environmental Affairs and Department of Water Affairs of the Botswana Government is acknowledged with gratitude. Many photographs were made available for the book by the photographers listed on page 144. Their generosity and interest in the project is appreciated.

Okavango Delta: Floods of Life was produced under the auspices of an agreement between IUCN and HOORC, and the *Biokavango* project of the Botswana government, the Global Environment Facility (GEF) and United Nations Development Program (UNDP); the UK DEFRA Darwin project *Monitoring and Simulating threats of aquatic biodiversity in the Okavango Delta* undertaken at the University College of London; and the Southern African Institute for Aquatic Biodiversity (SAIAB).

Recommendations to produce this book stem from a long-standing wish of staff at HOORC to have such a volume published and the IUCN project which produced the report *The Status and Distribution of Freshwater* ... *Southern Africa* published in 2009.

1

Introducing the Okavango Delta

The Delta of the Okavango River means different things to different people: a wonderland to some, a wetland or wildlife paradise to others, and for many it provides for wealth and welfare.

Few natural places in the world offer so many goods, assets and services in combination with such aesthetic appeal. And all of this happens in an incongruous setting lying in the centre of a flat, semi-arid landscape of sand that stretches almost three thousand kilometres from north to south, and over one thousand kilometres from west to east. Much of it is called the Kalahari Desert, and for its setting the Delta has been aptly named the jewel of the Kalahari, which is the title of another book that portrays the beauty and intricacy of the Delta.[1]

Several other books, and many magazine articles and documentary films have been devoted to its magic, their pages and footage filled with striking images of spectacular animals, plants and scenes. These images on film and paper have carried the fame of this natural asset out into the world. Each year, the Delta attracts tens of thousands of tourists from many countries on several continents.

Figure 1 | *David Livingstone and his co-explorers, William Cotton Oswell and Mungo Murray, were probably the first westerners to record their visits to the Okavango-Ngamiland region when they reached Lake Ngami in 1849.*[2] *Their 'discovery' was immediately followed by visits by many other explorers, hunters and traders, including those by Charles John Andersson and Thomas Baines. The earliest maps only documented features on the edges of the Delta, such as the Boteti and Thamalakane Rivers, Thaoge Channel and Khwebe Hills, and the Linyanti and Zambezi Rivers to the north. Although Livingstone had been told by local people that the country was 'full of rivers', it took explorers and cartographers several decades to comprehend the details of what lay inside the Delta. The map shown here was published in 1897 after a journey of exploration by Aurel Schultz and August Hammer. Theirs was the first map to show the Okavango River (given as* Okovanga*), and they inferred the rough courses of rivers through the Delta, naming them the Selinda and Tumulakane.*

While its margins became known to many visitors, the nature and extent of the Delta largely remained a mystery until 1905. This is when the explorer Siegfried Passarge published his Das Okawango-sumpfland und seine Bewöhner *(The Okavango swamps and its inhabitants). The book included the first map showing the approximate shape of the Delta and its network of waterways.*

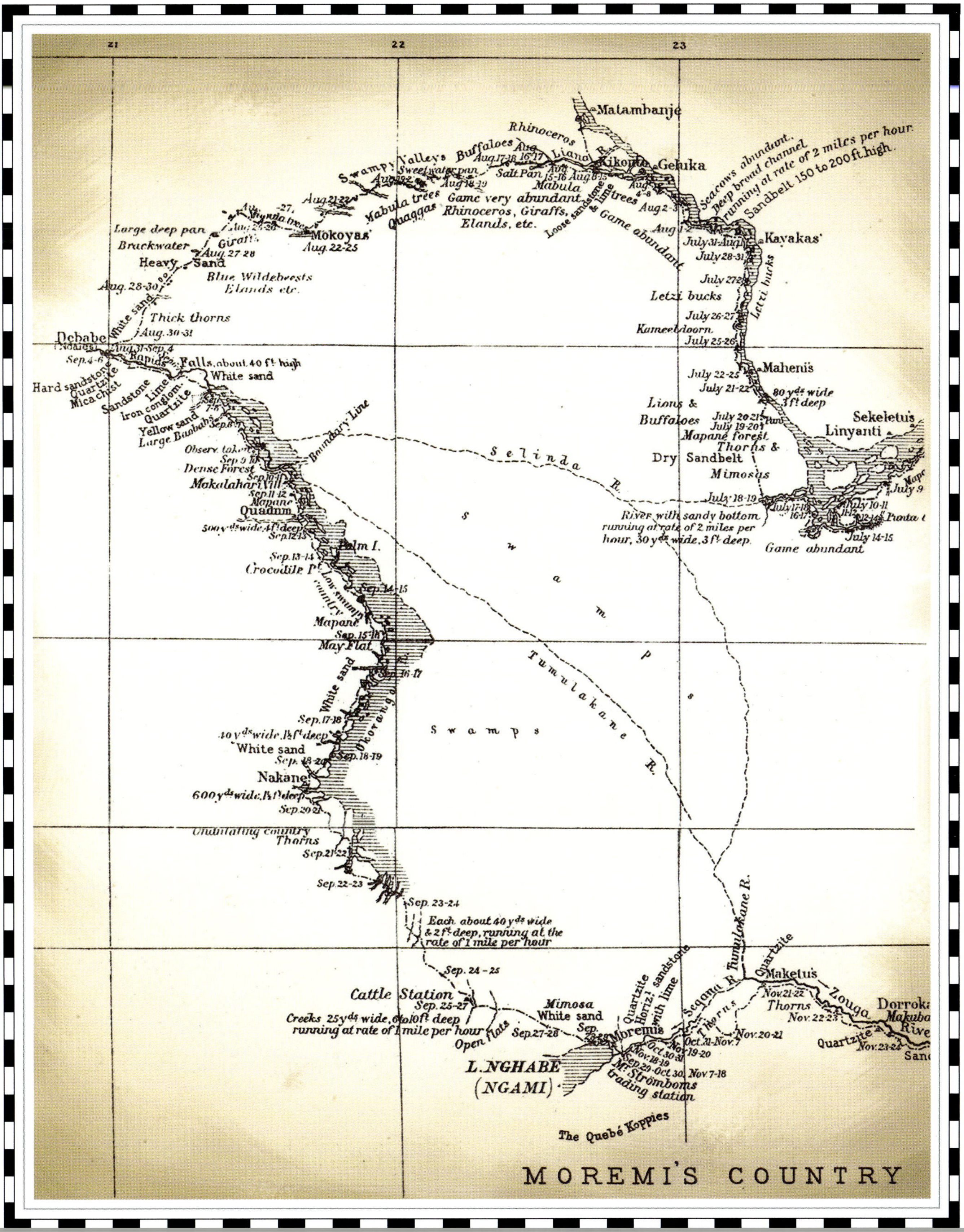
Matambanje
Rhinoceros
Swampy Valleys
Buffaloes
Sweetwater pan
Salt Pan
Mabula trees
Game very abundant
Rhinoceros, Giraffs, Elands, etc.
Quaggas
Mokoyas'
Aug. 22-25
Large deep pan
Brackwater
Heavy Sand
Blue Wildebeests
Elands etc.
Thick thorns
White sand
Debabe
Rapids
Falls, about 40 ft high
White sand
Hard sandstone
Quartzite
Mica chist
Sandstone
Lime
Iron conglom.
Quartzite
Yellow sand
Large Baobabs
Dense Forest
Makalahari Vill.
Quadum
Palm I.
Crocodile Pt
Mapane
May Flat
White sand
Nakane
Okovango R.
Boundary Line
Selinda R.
Swamps
Tumulakane R.
Kikonto
Geluka
Seacows abundant.
Deep broad channel
running at rate of 2 miles per hour.
Sandbelt 150 to 200 ft high.
Game abundant
Loose sandstone & lime
Kayakas'
Letzi bucks
Kameeldoorn
Mahenis
80 yds wide
3 ft deep
Lions & Buffaloes
Mapane forest
Thorns & Mimosas
Dry Sandbelt
Sekeletus
Linyanti
River with sandy bottom
running at rate of 2 miles per hour, 30 yds wide, 3 ft deep.
Game abundant
Each about 40 yds wide
& 2 ft deep, running at the rate of 1 mile per hour
Cattle Station
Creeks 25 yds wide, 6 to 10 ft deep
running at rate of 1 mile per hour
Open flats
Mimosa
White sand
Moremis
L. NGHABE
(NGAMI)
Mr. Strömboms trading station
The Quebé Koppies
Quartzite
Horiz. sandstone with lime
Seqana R.
Tumulakane R.
Thorns
Maketus
Zouga
Dorroka
Makuba
River
Quartzite
MOREMI'S COUNTRY

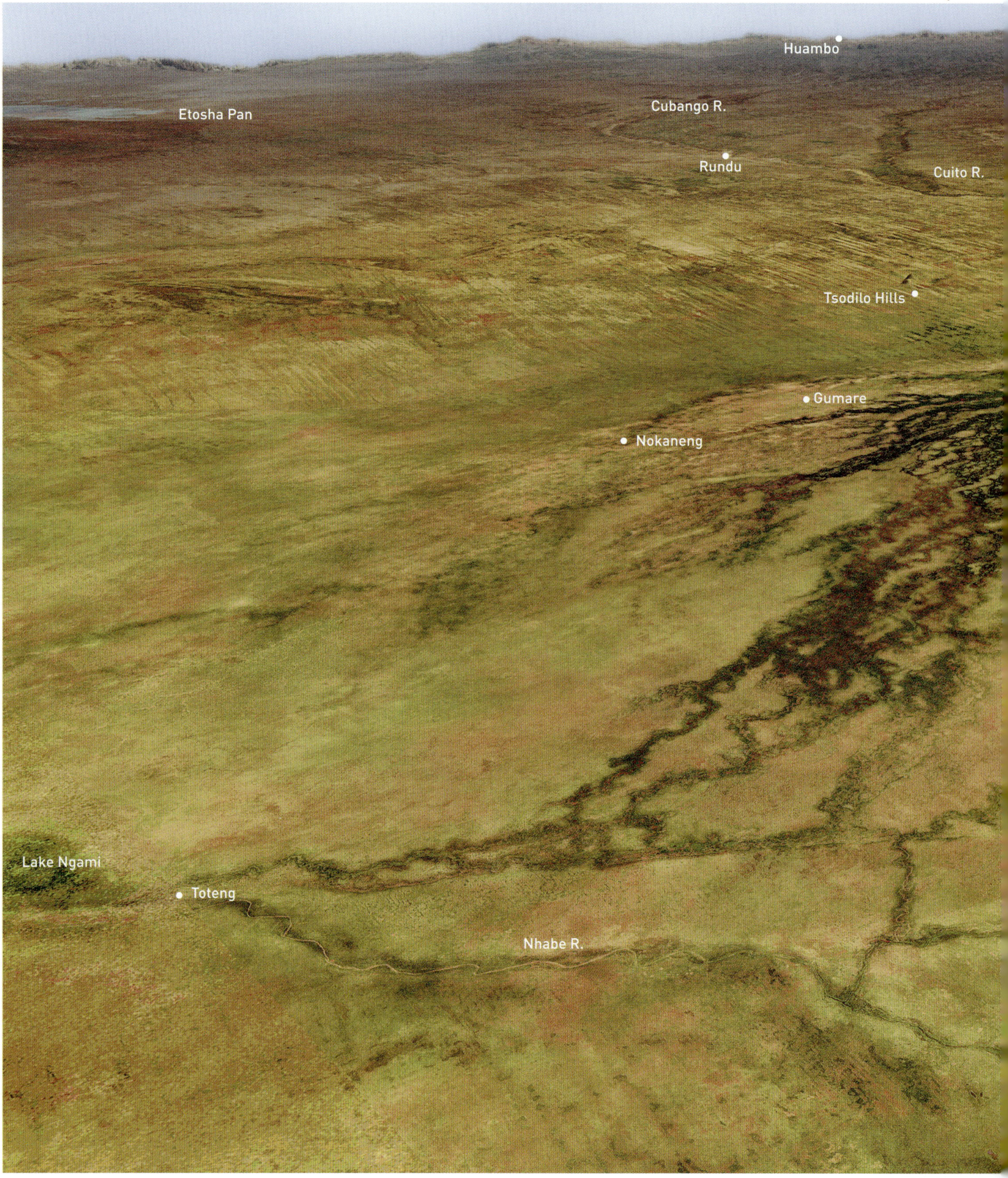
Huambo
Etosha Pan
Cubango R.
Rundu
Cuito R.
Tsodilo Hills
Gumare
Nokaneng
Lake Ngami
Toteng
Nhabe R.

Figure 2 | *The Delta's alluvial fan is in the foreground, the Panhandle behind it, while the Angolan catchment commands the background. Panhandles are narrow strips protruding from larger areas of land, while deltas and alluvial fans are defined as triangular or fan-shaped plains of alluvial sediments deposited where the gradients of rivers suddenly decrease.*[3] *This is usually at the mouth of a river where it meets the sea – and that is called a delta – or where a river issues onto a broad plain, where the sediments would be an alluvial fan. The latter definition applies more closely, but calling this the Okavango Alluvial Fan would be indeed cumbersome!*

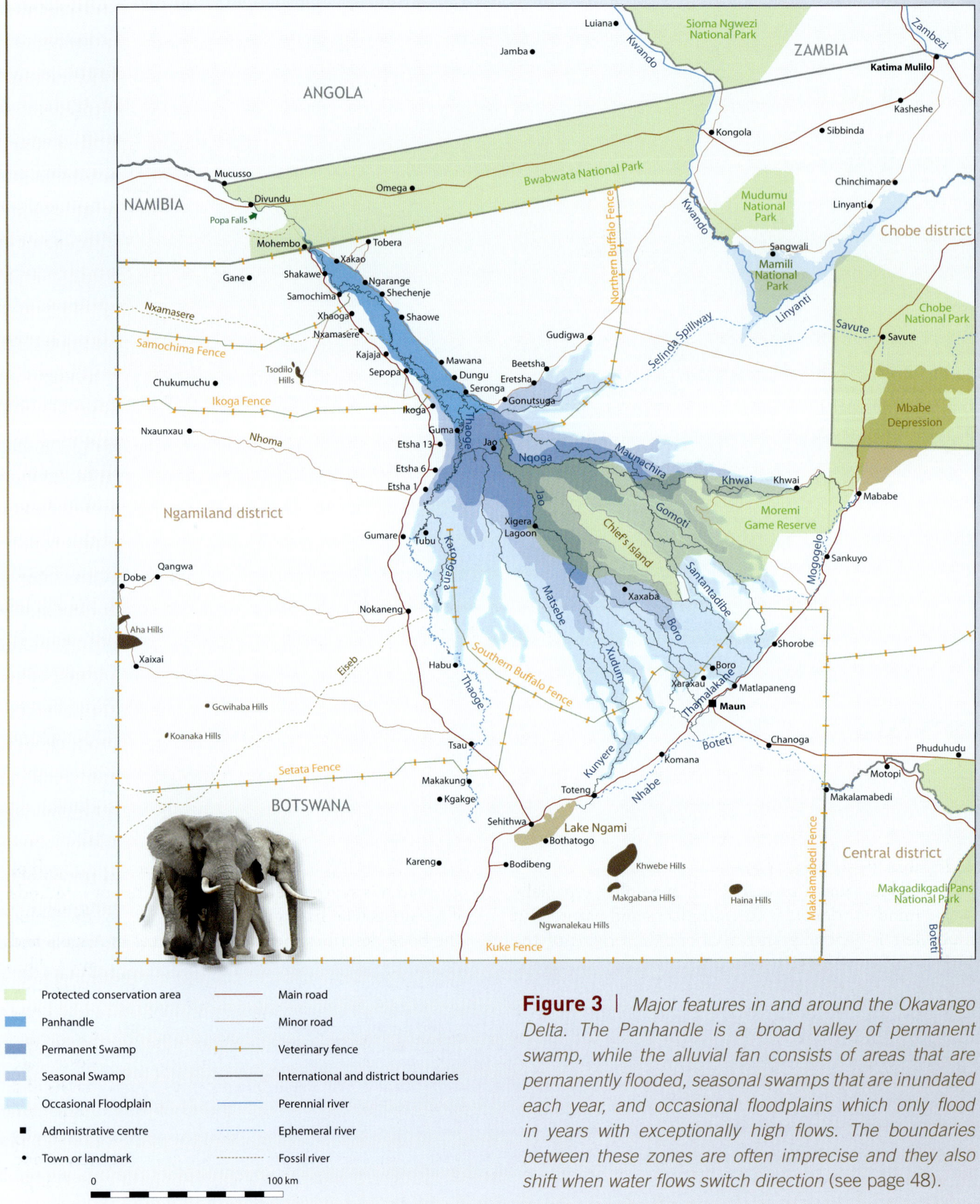

Figure 3 | *Major features in and around the Okavango Delta. The Panhandle is a broad valley of permanent swamp, while the alluvial fan consists of areas that are permanently flooded, seasonal swamps that are inundated each year, and occasional floodplains which only flood in years with exceptionally high flows. The boundaries between these zones are often imprecise and they also shift when water flows switch direction* (see page 48).

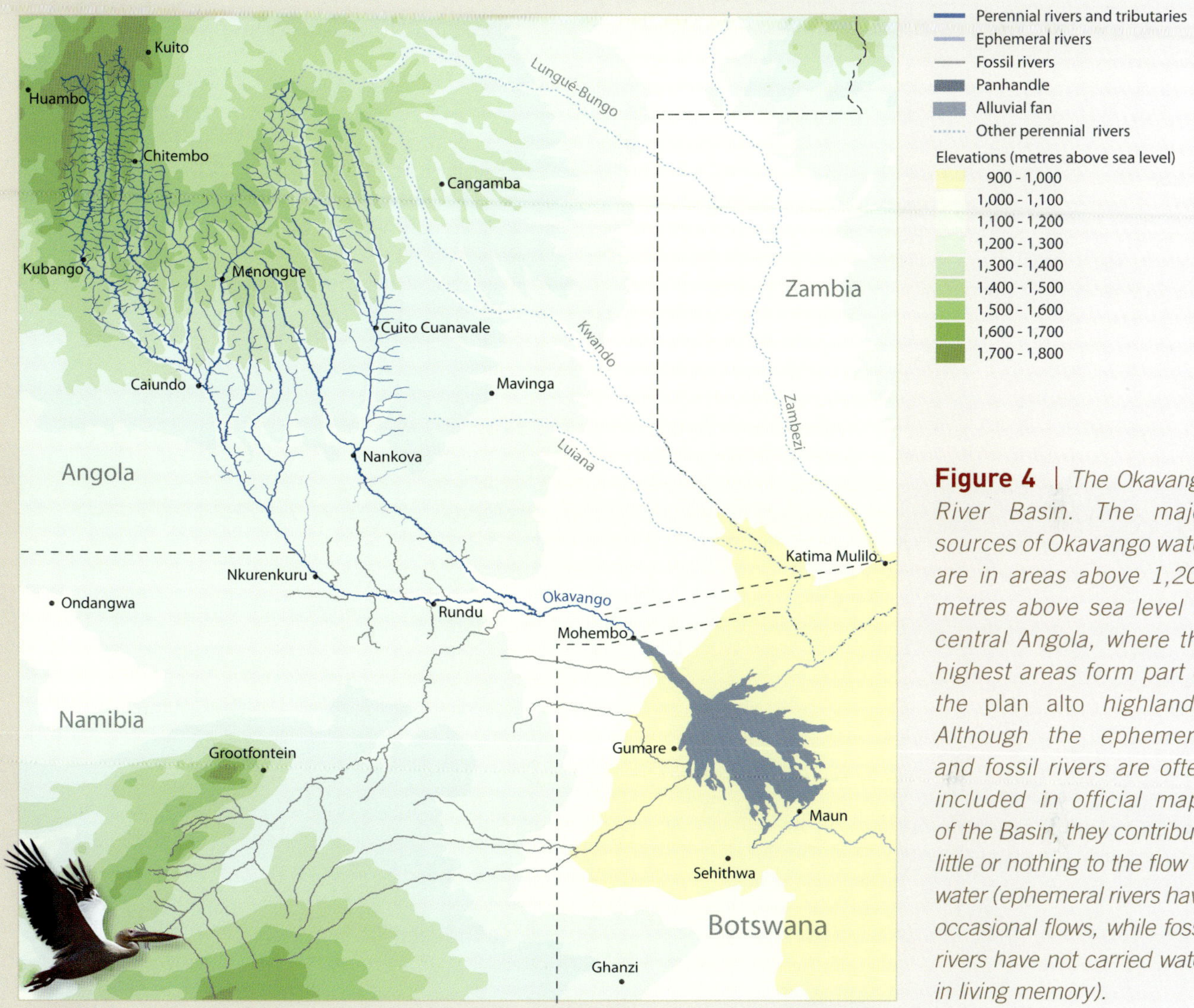

Figure 4 | *The Okavango River Basin. The major sources of Okavango water are in areas above 1,200 metres above sea level in central Angola, where the highest areas form part of the plan alto highlands. Although the ephemeral and fossil rivers are often included in official maps of the Basin, they contribute little or nothing to the flow of water (ephemeral rivers have occasional flows, while fossil rivers have not carried water in living memory).*

Compared to other freshwater environments, the Delta is thus well known as a spectacle rich in wildlife, especially in the minds of devotees to wild places and nature. But much less is generally known about the origins of the Delta, its functioning, the major processes that give it bountiful life, what threats it may face in the years ahead, and what can be done to meet challenges that may compromise its future. And limited information about its workings is available to people who may influence that future, for example those who determine how water in the Delta's catchment is used or abused.

Okavango Delta: Floods of Life is a celebration of the

The first is to explore how the Delta functions. Much of our search is synthesised from the findings of hundreds of scientists who have given years of study to the Delta. The book describes the catchment areas in Angola and the passage of the Okavango River through Namibia. This is where the quality, volume and timing of water flow entering the Delta is determined. Once water reaches the Delta, various processes govern the distribution of flows across the alluvial fan. Some processes are physical, dependant on slope, sedimentation, faults and channels fixed long ago in the sands of the Kalahari. Yet others are driven by biological agents, principally by such divergent organisms as papyrus, termites and

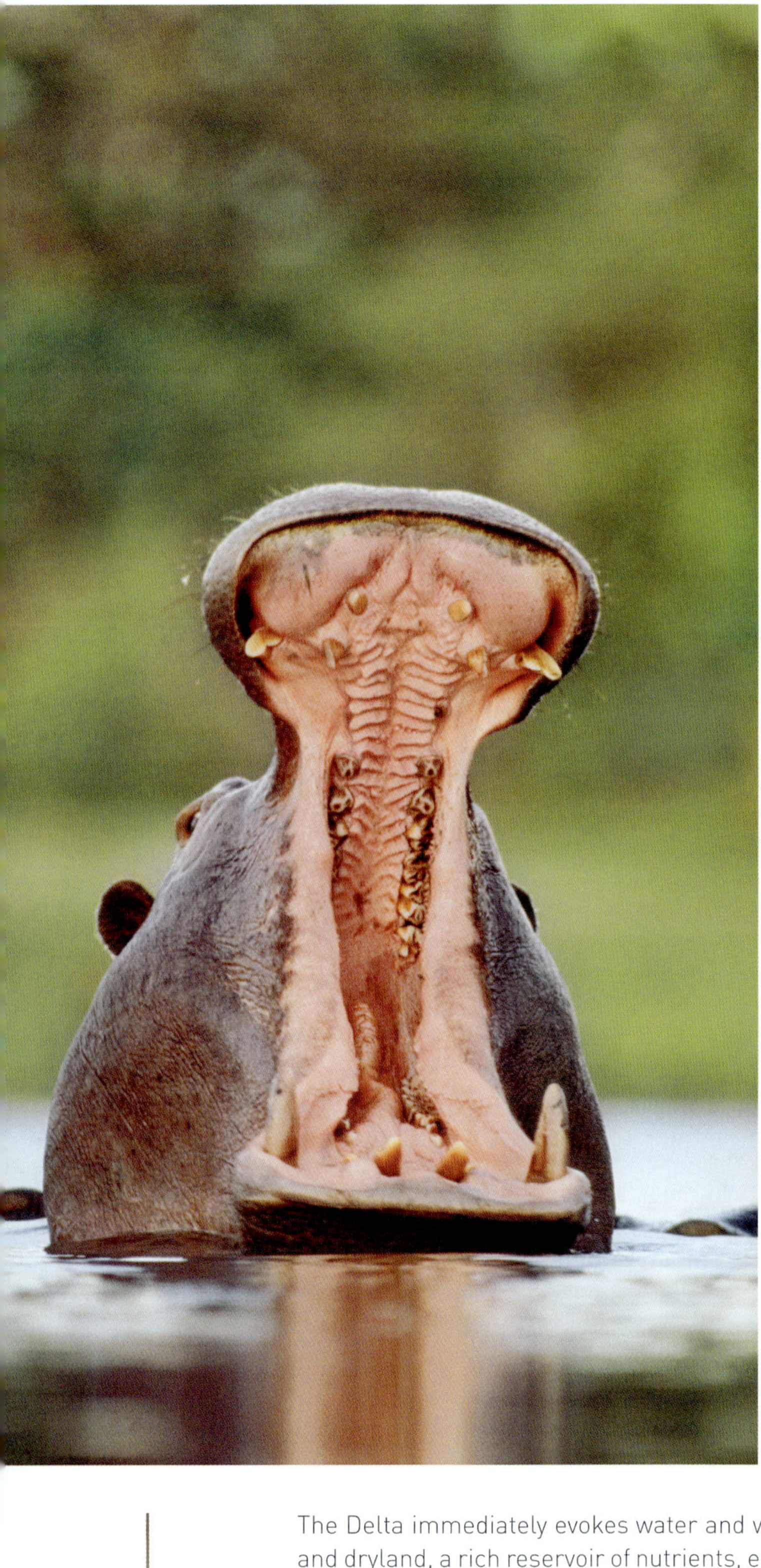

The Delta immediately evokes water and wildlife. But there is so much more: a meeting between wetland and dryland, a rich reservoir of nutrients, engineering processes that move water from one place to another and keep the water fresh; a patchwork of diversity of habitat and life forms; three countries, each with separate perspectives and influences on the Okavango River and its Delta; high economic values that move people of the Delta from subsistence lifestyles to ones enmeshed in cash-based services; one of the world's most pristine river systems; and tangible proof of good governance.

next, and across longer scales of time that range from tens to thousands of years. In short, the flow of water is extremely variable and unpredictable.

Pulses of water coming down from the catchment are critical to the functioning of the Delta. But the rich production of plants and animals is only possible when – and where – the waters mix with layers of nutrients to mobilise these molecular constituents of food and form. Without abundant nutrients, the waters of the Delta would be sterile, and so it is this dynamic mix of water and nutrients, ever shifting from one place to another, that holds the key to the wealth of biological production and diversity.

A second aim of this book is to emphasize the values of the Delta that stem from the rich life found here. This wealth has been recognized by the international Ramsar Convention, which has distinguished the Delta as a wetland of 'significant value for Botswana and ... for humanity as a whole ... because of its international significance in terms of ecology, botany, zoology, limnology or hydrology'.[4]

Not only is the Delta an oasis of water, but also one of high biological diversity with more than 400 species of birds, about 1,300 plants, 71 fish and tens of thousands of invertebrate species, for example. What is significant about these organisms? Where else in the world are they found? On what habitats do they depend, and what factors determine their abundance? What assets, resources, products and services are provided by the Delta's biota, and what or whom do they benefit?

What needs to be done to secure the future of the Delta? To address this question is the third purpose of *Okavango Delta: Floods of Life*. Although this Kalahari wetland is isolated from other expanses of freshwater, it is by no means immune to various external threats. There are also local challenges brought about by economic activities that take place in and around the Delta. Most of these relate to the way in which the people of the Delta live and regard its resources.

A range of approaches and activities need to be employed if the Delta is to continue to serve its people, the citizens of Botswana and the international community. Reliable information is needed to assess the nature and

The elevated mounds of termites later become wooded islands which are crucial in adding to the diversity of habitats in the Delta and maintaining its fresh water.

increasing, it is important to monitor the health of the Delta so that detrimental impacts can be identified as early as possible.

Such technical assessments and measurements are necessary, but alone they are not sufficient. The challenges facing the Delta derive ultimately and proximally from economic interests, which are usually cloaked in the languages of politics and business. The choice for conservation thus needs to be backed by political will.

All who have an interest in the Okavango Delta need a common purpose and tongue if they are to preserve this valuable oasis of water, wildlife, wealth and welfare. We

Across a fire-scarred Panhandle to the woodlands that line its western edge, and to the Tsodilo Hills in the distance.

2

Shaping the Delta

Whereas the great majority of rivers in the world flow to the sea, the Okavango River ends in its Delta, right in the middle of the southern African continental landmass.

The Delta is close to the lowest point and centre of the Kalahari Basin (**Figure 5**). Most of the basement rocks beneath the Kalahari and the Delta were formed long ago from sediments that were forced and folded upwards from the floor of an ancient ocean about 700–550 million years ago. Extreme pressure and heat turned the marine sediments into metamorphic rocks of the so-called Damara Group. Overlying the Damara Group rocks are more recent fluvial, aeolian and volcanic rocks of the Karoo Group deposited from 300–180 million years ago. All these basement rocks are now hidden beneath the surface, except as isolated landmarks, such as the Khwebe Hills south of Toteng and the Tsodilo Hills west of Sepopa (*see* **Figure 3**, *page 14*).

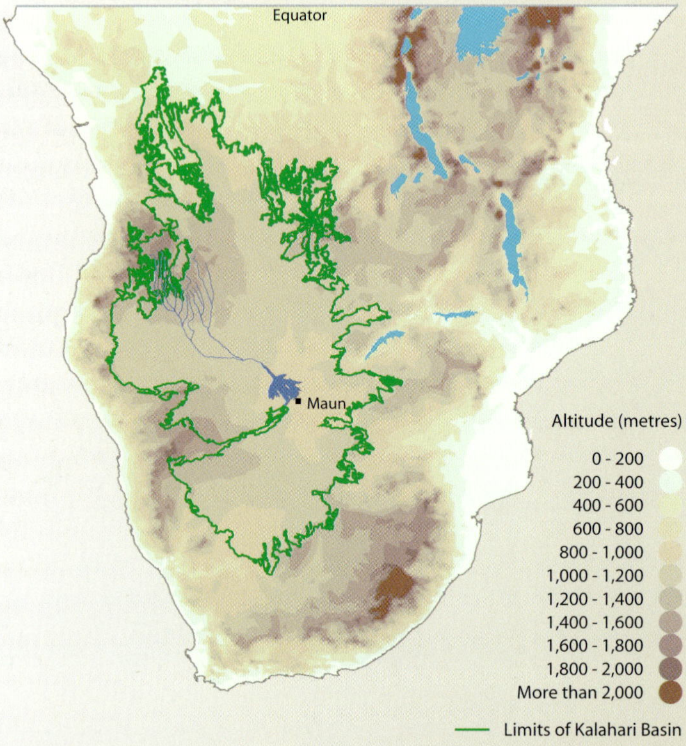

Figure 5 | *The Kalahari Basin is a vast sea of sand extending 3,000 kilometres from the northern Cape, across most of Botswana, through much of Angola and up to the Democratic Republic of Congo. From west to east, the Basin extends 1,500 kilometres at its widest.*

Two sets of faults in the basement have a marked impact on the flow and distribution of surface rivers and the shape of the Delta. Most of the first group are far to the north, where a series of parallel faults direct the flows of the Okavango, Cuito, Kwando and Zambezi rivers south-eastwards (*see* **Figure 4**, *page 15*). Extensions of these faults probably fix the orientation of the Panhandle, as well as the south-easterly flow of the Kwando River before it kinks into the Linyanti Swamps.

The second set of faults lies perpendicular to the first, and these control the position and shape of the Delta. The faults were formed by tectonic activity along a south-western extension of the East African Rift Valley. About two million years ago, faulting in this extension probably created two sub-basins in northern Botswana. One of these lies beneath the Delta, while the other sub-basin is found below the Makgadikgadi Pans. It is probably that rifting and collapse of the underlying basement that caused this broad trough, which encompasses the Delta and Makgadikgadi to be the lowest in the whole Kalahari Basin.

Other faulting along this extension of the Rift Valley is visible on the surface as the Gumare, Kunyere, Thamalakane, Linyanti and Chobe faults. The Gumare fault divides the Panhandle from the alluvial fan. The Kunyere and Thamalakane faults block water in the Delta from flowing further south-eastwards (**Figure 7**), forcing it to slow and therefore to spread. These two faults are fundamental in confining the Delta's alluvial fan.

The ridges that form the margins of the Panhandle rise about 20 metres above the swamp (**Figure 6**), and are sometimes thought to have been formed by faulting. However, it appears likely that the broad valley that confines the Panhandle was caused by erosion as the Okavango cut through, and deeper into the surrounding Kalahari Sands.

The Kwando/Linyanti has been shaped in the same way as the Delta, its flow to the south-east being halted by the Linyanti fault. As a result, the Kwando forms a mini-delta. Flows seep away progressively through the Linyanti Swamps to the

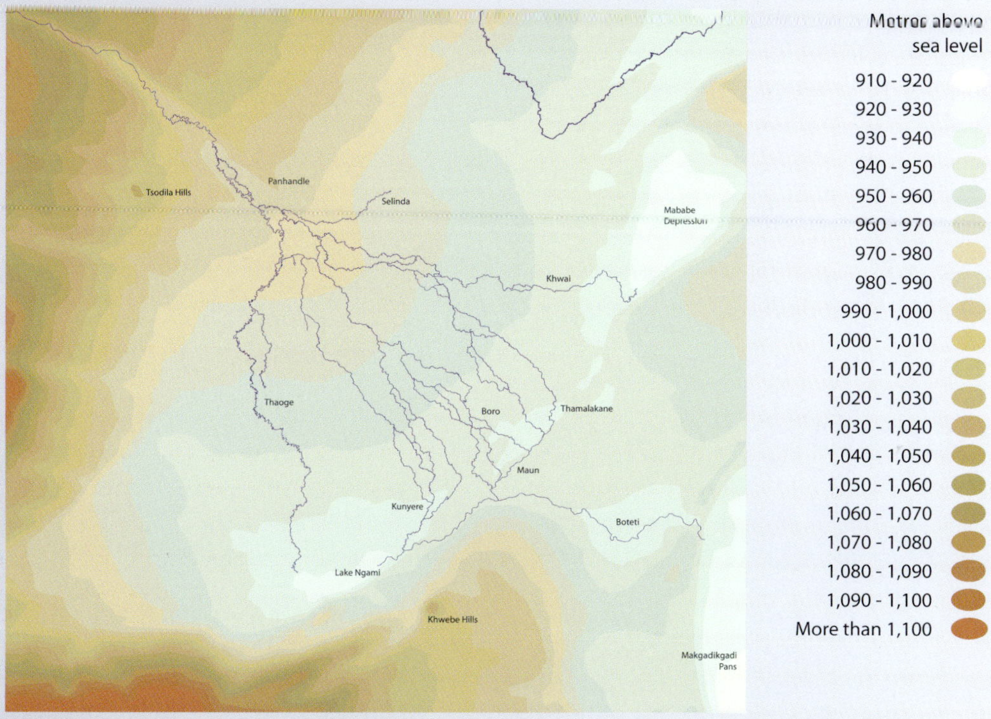

Figure 6 | *Although slopes down and across the Delta are very even, the deposition of sediments bought in by the Okavango River has led to the formation of a fan with a near-perfect conical shape, as shown in this map of elevations above sea level.*

BELOW: Popa Falls marks the real head of the Delta. As water tumbles over the Falls which are a nick-point in the Okavango River, its speed of flow slows and it is there that the river widens into the Panhandle and begins to deposit sandy sediments.

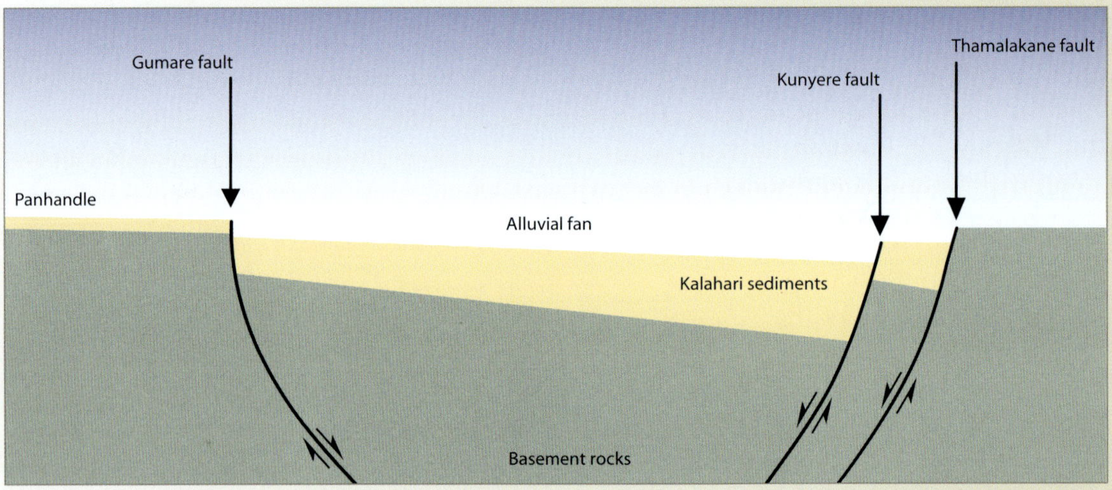

Figure 7 | *Cross-section through the Delta from north-west to south-east, showing the layers of rock and sediments, and the three faults that confine the alluvial fan.*[1] *Elevations along the Panhandle are extremely gentle, the river dropping only about 5 metres over a distance of about 150 kilometres. Over the next 150 kilometres or so from the Gumare to the Thamalakane Fault at the edge of the alluvial fan, elevations drop about 40 metres.*

Figure 8 | *The surface geology in and around the Delta includes well-developed sand dunes which formed during very arid conditions, as well as massive areas of alluvial sediments deposited during much wetter times by the Okavango, Kwando and Zambezi rivers. Recent earthquakes have ranged in magnitude from 2 (the smallest dots) to 6.7 on the Richter scale.*[2]

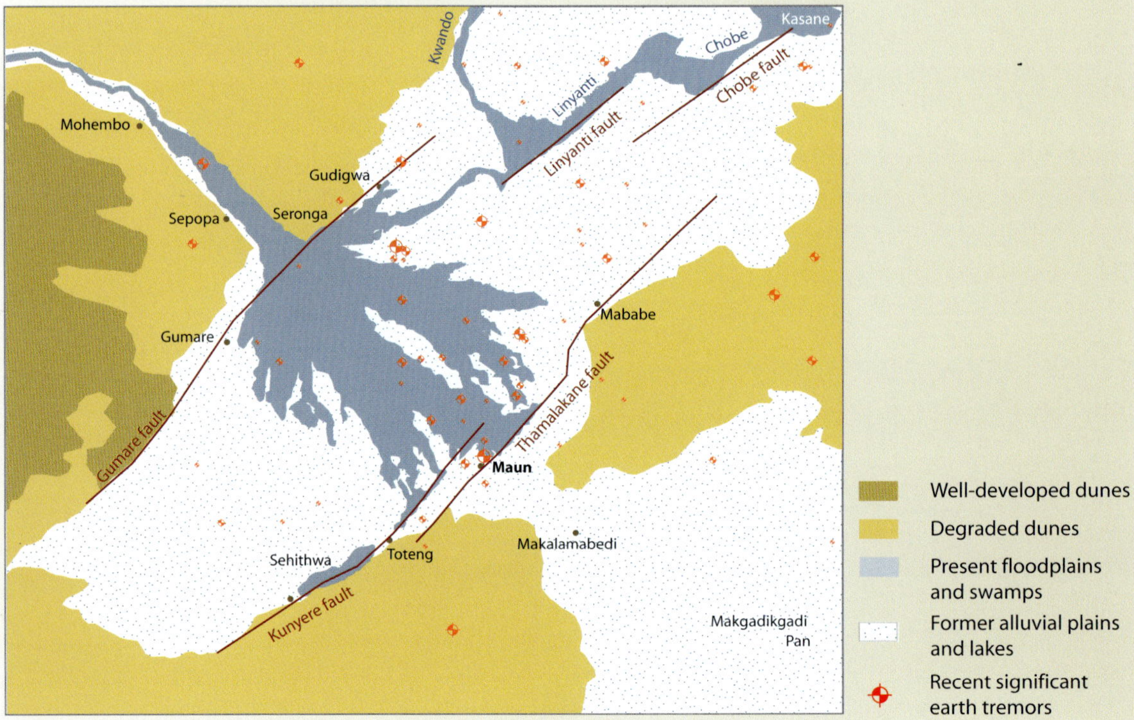

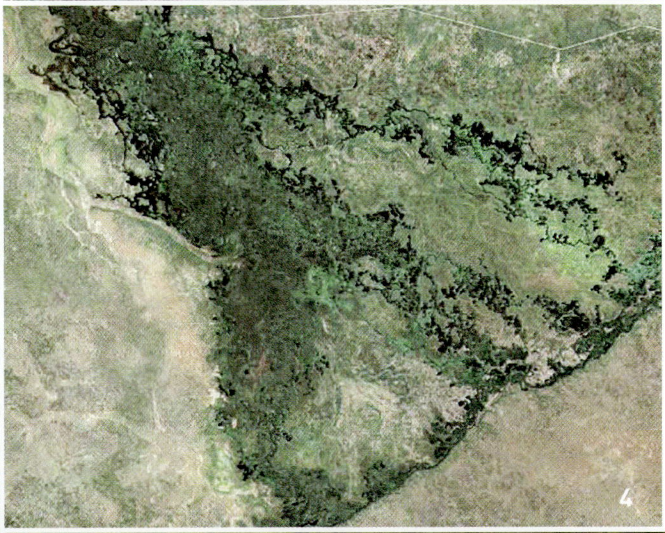

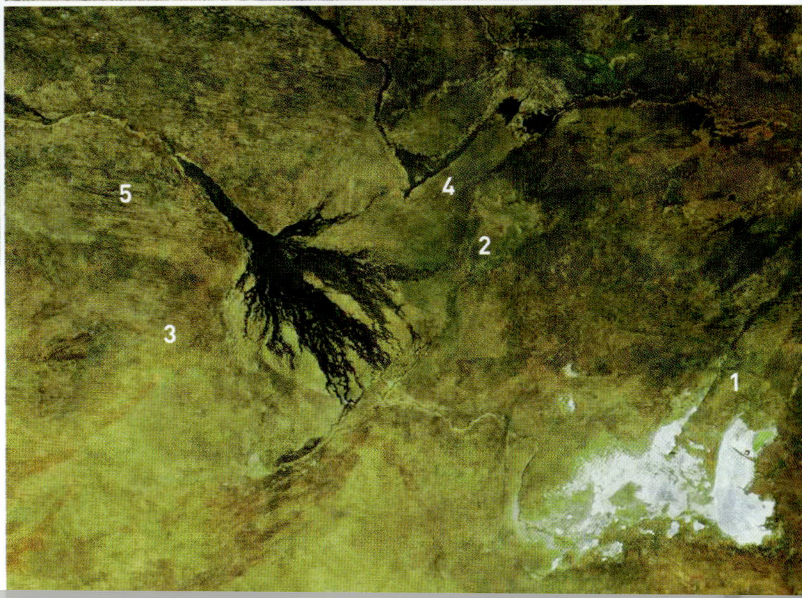

Some features around the Delta:[3]

1. The salt works in Sua Pan, which is part of the great Makgadikgadi Pans.
2. Mababe Depression; the colours reflect scars from fires at different times.
3. The sharp interface between the sand dunes and alluvial deposits of the Delta when it was much larger.
4. The Kwando River where it forms a mini-delta with the Linyanti Fault.

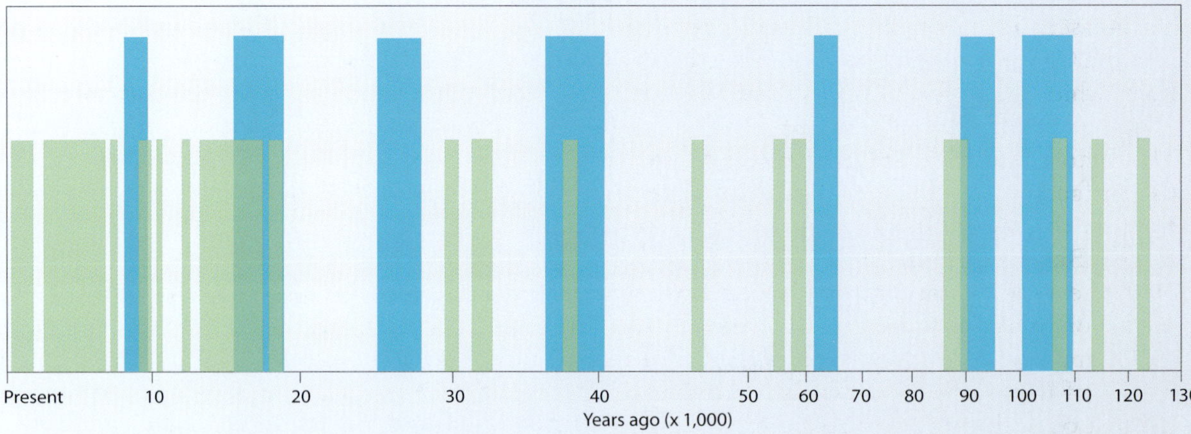

Figure 9 | Top: *Periods (blue bars) during the last 130,000 years ago when mega-lakes extended over Makgadikgadi and the low-lying areas around Lake Ngami, the Mababe Depression and across the eastern Caprivi, and phases (green bars) when Lake Ngami and its immediate surrounds were flooded. The extent of the mega-lakes (which covered some 66,000 square kilometres) and flooded areas around Lake Ngami (covering about 2,600 square kilometres) are shown in the map below. The map also depicts the extent of alluvial sediments deposited during other wet periods and when the flows of the Okavango, Kwando and Zambezi Rivers probably followed different courses.*[4]

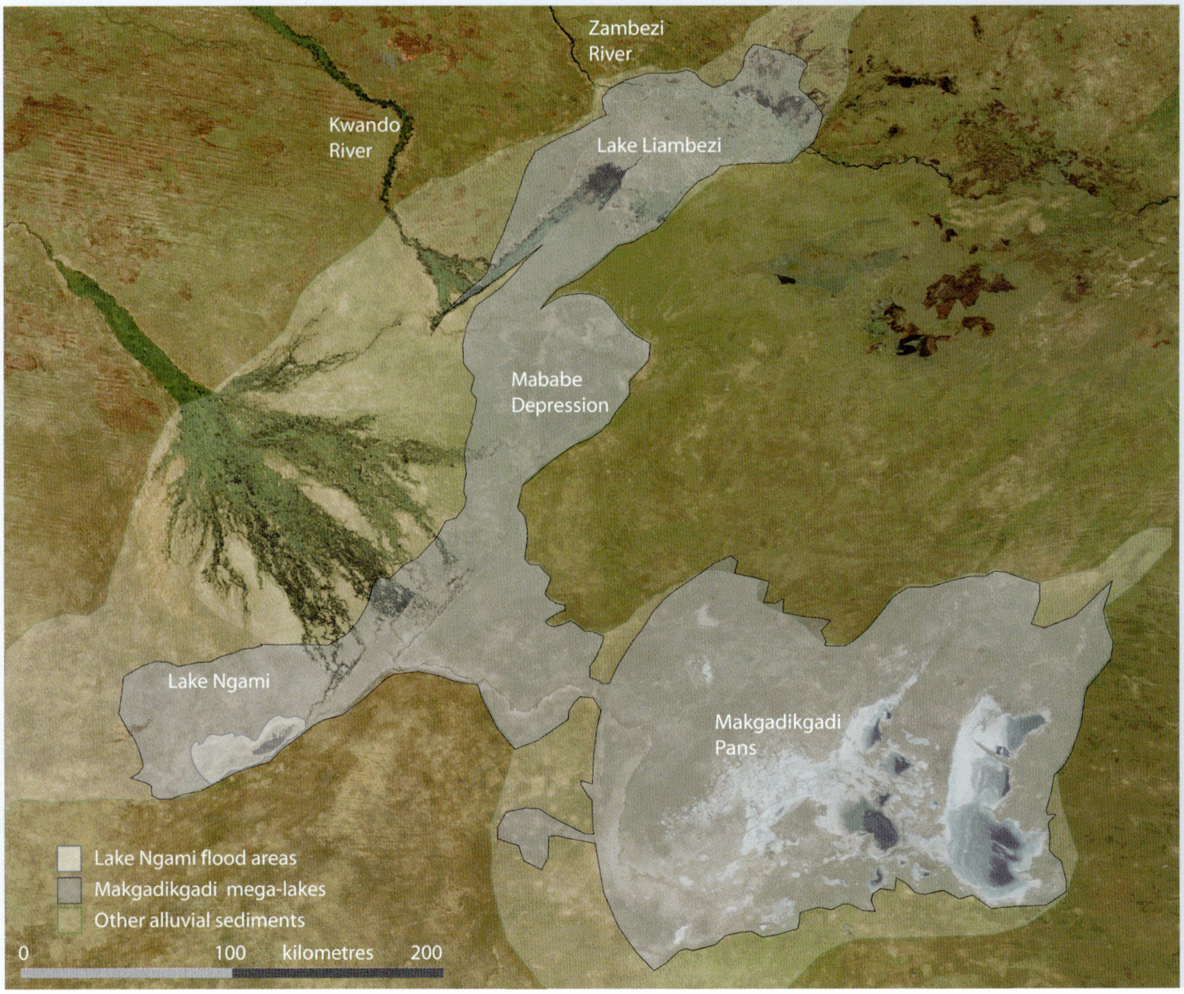

north-east and sometimes reach Lake Liambezi. Further east, the Chobe fault controls the position of the Chobe River, which is a backwater of the Zambezi River.

The whole Kalahari Basin now has a remarkably flat surface, which drops gradually from its margins towards the centre. Its flat surface is a consequence of the Basin being filled steadily with sediments over the past 65 million years. Several hundred metres of layered sediments are to be found in places. For example, the Delta lies on top of 100 to 270 metres of deposits.[5] The sediments laid down during the first 63 million years were mainly alluvial, carried into the Basin by large rivers that left their deposits in massive lakes and deltas. The remnants of these now lie deep below the present surface, which, by contrast, was largely shaped by aeolian deposits of wind-blown sand during much more arid conditions over most of the last two million years.

But even during these drier two million years, other rivers bigger than the Okavango have flowed into northern Botswana, leaving behind alluvial sediments over an expanse of about 120,000 square kilometres around the Delta and Makgadikgadi Pan **(Figure 9)**. That area is about ten times bigger than the present Delta. For example, the Zambezi possibly ran here as recently as 50,000 years ago, before being captured by a river that had cut its way back from the east coast.[6] An older course of the Kwando also probably flowed into what is now the Delta.[7]

The Okavango Delta and northern Botswana have therefore looked very different from what we see today. At times dominated by aridity, these areas may have been very similar to the Namib Desert in appearance, while in wetter periods huge expanses of water in lakes, pans and perhaps other deltas have characterized the landscape.

How old is the Delta? This is difficult to answer, given the variety of rivers, courses and volumes of water that have flowed into northern Botswana over millennia. However, if we define the Delta as the fan of water now confined by the Kunyere and Thamalakane faults, it is perhaps relatively young, since those barriers seem to have developed no more than 120,000 years ago. We presume this because extensive older alluvial

In the same way that climates change, tiny tectonic movements may cause subtle changes in elevation which result in significant changes in the distribution of flows across the Delta (*see page 51*). Most of the dozens of tremors in the area have been small **(Figure 8)** but one at Maun in 1952 measured 6.7 on the Richter scale.

The repeated flooding and drying of the Delta and its adjacent areas over hundreds, even thousands of years may be considered a matter of academic interest. However, that history and the confinement of flows has led to the accumulation of nutrient resources in and around the Delta. Each flood brought with it more of these chemical constituents of life. Nutrients blown in by wind were trapped in the water, and faecal matter from animals attracted to the water and its associated floodplain pastures likewise enriched the Delta. The key point is that the rich nutrient supplies took a very long time to accumulate here, unlike the fresh supplies of water which arrive to re-flood the Delta year after year. Maintaining the productivity of the Delta thus requires the protection of both the supply of water and the age-old base of nutrients.

KEY POINTS

1. The Delta lies near the lowest part of the vast Kalahari Basin which stretches over 3,000 kilometres from north to south and up to 1,500 kilometres east to west.
2. The Basin is largely filled with aeolian and alluvial sediments that lie on basement rock formations.
3. Faults lying north-west to south-east in the basement rocks control the flow of the Okavango River south-eastwards towards the Delta.
4. Other perpendicular faults – the Gumare, Kunyere and Thamalakane faults – confine the alluvial fan of the Delta.
5. Climates and the extent of the Delta and other bodies of water in northern Botswana have changed radically. There have also been dry periods when the Delta was probably covered by sand dunes.
6. As a result of flooding, nutrients have accumulated progressively over a long time in the Delta's sediments and those of adjacent

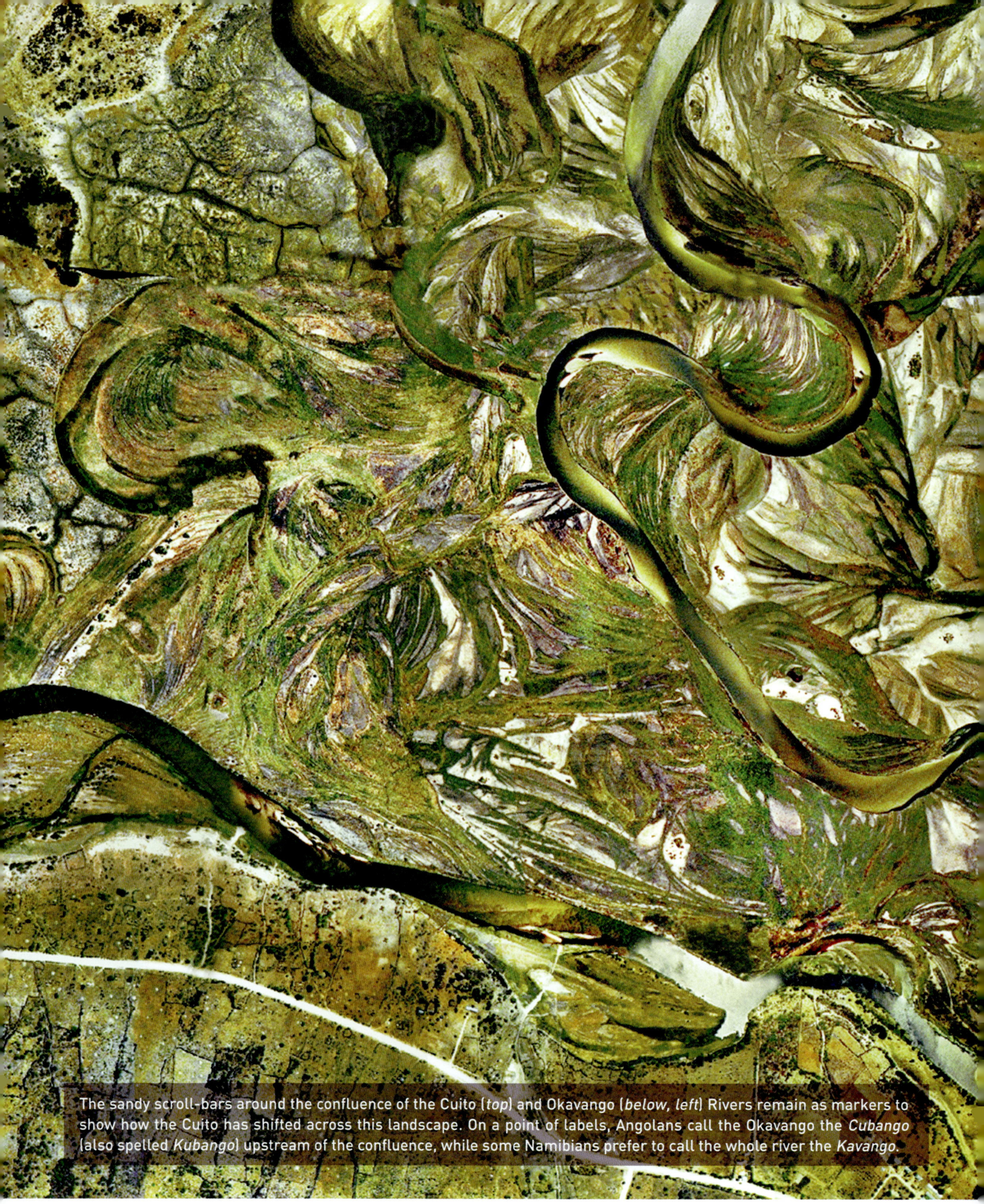

The sandy scroll-bars around the confluence of the Cuito (*top*) and Okavango (*below, left*) Rivers remain as markers to show how the Cuito has shifted across this landscape. On a point of labels, Angolans call the Okavango the *Cubango* (also spelled *Kubango*) upstream of the confluence, while some Namibians prefer to call the whole river the *Kavango*.

3 Okavango River: flow of a lifeline

For most people, the word Okavango brings the Delta to mind, a place given prominence by the public media and tourism, and renowned for its rich wildlife.

However, there is another part to the Okavango that should enjoy equal fame. This is the Okavango River's catchment, much of which consists of tens of thousands of square kilometres of the most pristine wilderness in an area once appropriately known in Angola as *terras do fim do mundo* – the land at the end of the earth. Moreover, the wealth of the Delta depends largely on the health of the catchment. The future of the Okavango Delta is thus rooted in the future of the Okavango River!

The whole Okavango River Basin is divided into three zones: the northern catchment from which water actively drains, the middle-reaches through which the Okavango River flows without collecting significant runoff, and then the Delta **(Figure 10)**. The northern area lies entirely in Angola and receives much higher rainfall than the middle zone, which is in southern Angola and Namibia. In addition to the differences in rainfall, there are also substantial changes in the use of land, demography and recent history across the whole Basin. All these aspects have direct or indirect bearings on the Delta.

The northern catchment spans an area of about 400 kilometres from east to west and 500 kilometres from north to south, and is divided into two sub-catchments. The western, or Cubango sub-catchment is relatively hilly and characterised

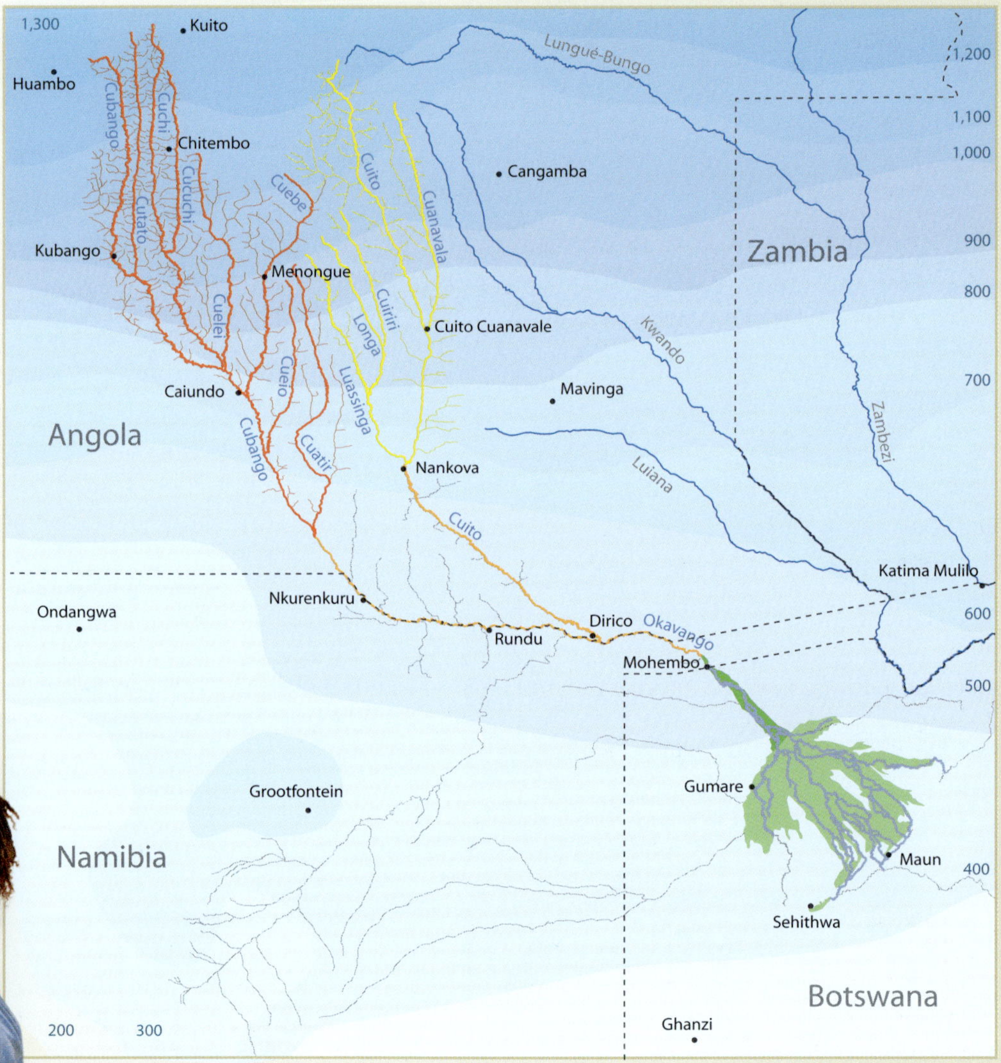

Figure 10 | *The major zones and catchment areas that make up the Okavango River Basin, and average annual rainfall in and around the Basin.*[1]

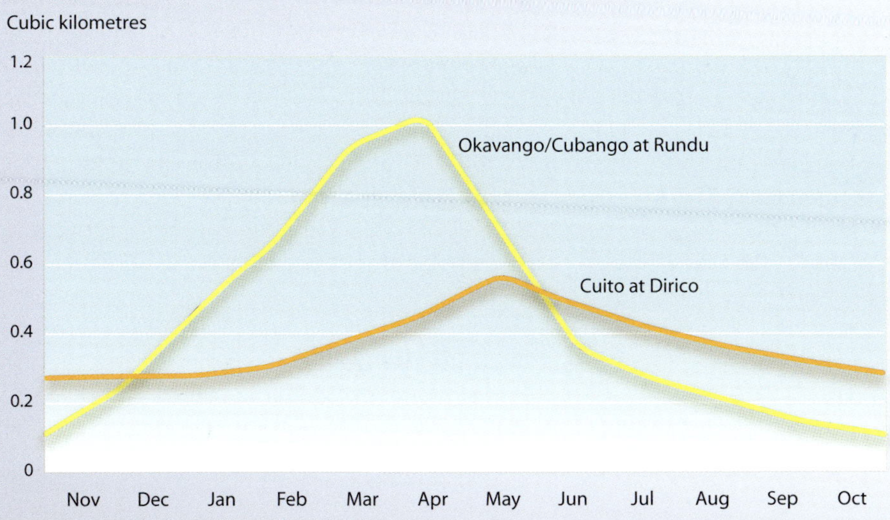

Figure 11 | *Flows down the Cubango sub-basin are rapid, peaking at the end of the summer rains, while the slower flows down the Cuito only reach its confluence with the Cubango/Okavango two or three months later. The figures are average volumes of river water passing Rundu (above the Cuito confluence) and Dirico just before the Cuito joins the Okavango.*

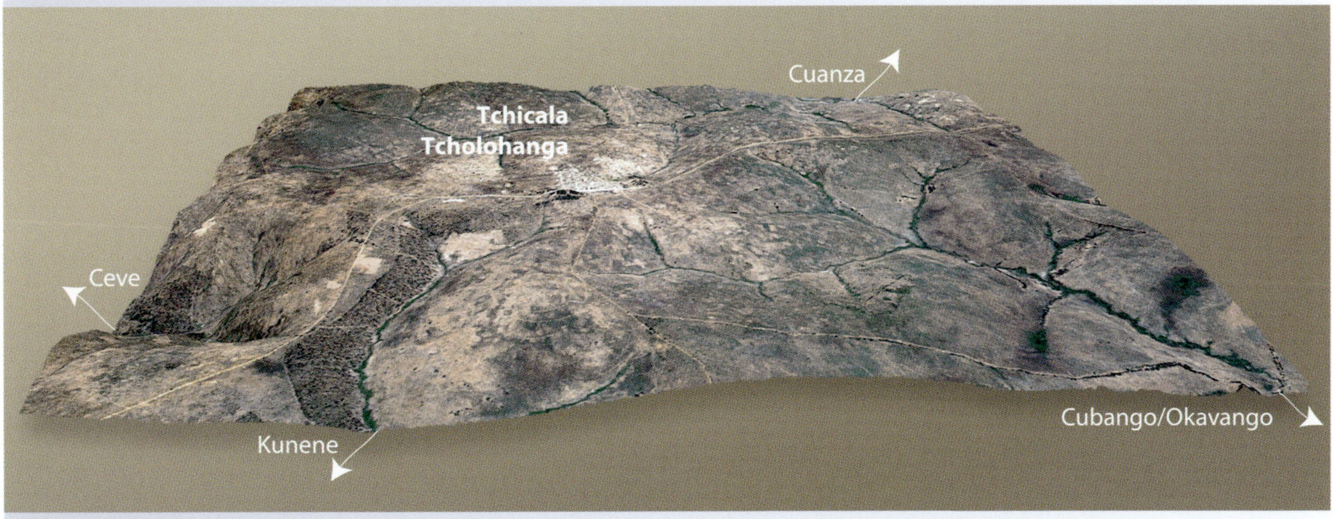

The Okavango River starts its journey to Delta just outside the small town of Tchicala Tcholohanga on the *plan alto* (highlands) of Huambo Province, Angola. Remarkably, the small hill where the town was built is also the source of three other major rivers. These, and the Okavango flow in quite separate directions: the Kunene River to the south-west and its eventual mouth in the Atlantic Ocean, the Ceve north-west to its mouth at Port Amboim on the Atlantic, the Okavango south-east to the Delta, and the Cutato (a major tributary of the Cuanza) north-east and then westwards to the Atlantic just south of Luanda. The river basins associated with the four tiny tributaries fanning out from small hill at Tchicala Tcholohanga thus cover hundreds of thousands of square kilometers of Angola, Namibia and Botswana.

by outcrops of granites, ferasol soils and Kalahari sands, while the eastern, Cuito sub-catchment consists almost entirely of Kalahari sands that blanket a flat landscape.

Those flows coming off the western, higher areas are more rapid and aggressive, water running downstream soon after rain has fallen. Sharper increases in inflow to the Delta are therefore mainly of water coming from the

Images of Angola: Many rivers are flanked by wide grassy floodplains (*top left*), while their meanders are often so broad that the crystal-clear waters flow sideways as much as they do downstream (*bottom left*). Hundreds of kilometers of valley marshlands are dominated by papyrus which filters out nutrients, clays and sandy sediments (*top right*). Maize, cassava, bananas and various vegetables are grown in *onaka* fields along tiny tributaries (2nd *from top right*), while rivers provide the only water for most villagers to wash their clothes (3rd *from top right*). Remnants of the civil war, which ended in 2002, are scattered across the catchment (*bottom right*).

Images of Namibia: The well-wooded islands at Andara have not been cleared because they are the traditional burial grounds of Hambukushu chiefs (*top left*), but most riverine vegetation has been cleared for crops elsewhere (*middle left*). The Okavango flows along a narrow, steep-sided valley in some stretches, while elsewhere it meanders across broad floodplains (bottom left). Most tourist lodges (*top right*) are near Rundu or within 50 kilometres of Mohembo. Goats and cattle are much more abundant than in the Angolan catchment (*middle right*). The Namibian government intends to expand the current 2,200 hectares of irrigation to a potential area of 15,600 hectares.[2]

By contrast, flows from the Cuito sub-catchment provide a more sustained input of water to the Delta, since much of the water entering the Cuito and its major tributaries – the Longa, Cuanavale and Luassinga Rivers – is slow seepage from the surrounding Kalahari sands. Moreover, the Cuito and its tributaries meander across broad valleys that have only the slightest gradient. For example, the Cuito River drops just 85 metres over a straight-line distance of about 270 kilometres from Cuito Cuanavale to Dirico. And the flow of water is further slowed as it percolates through the broad floodplains of grasses, reeds and sedges that line the main channels of these rivers.

Figure 12 | *The density of people throughout the Basin and in Angola, Namibia and Botswana.*[3]

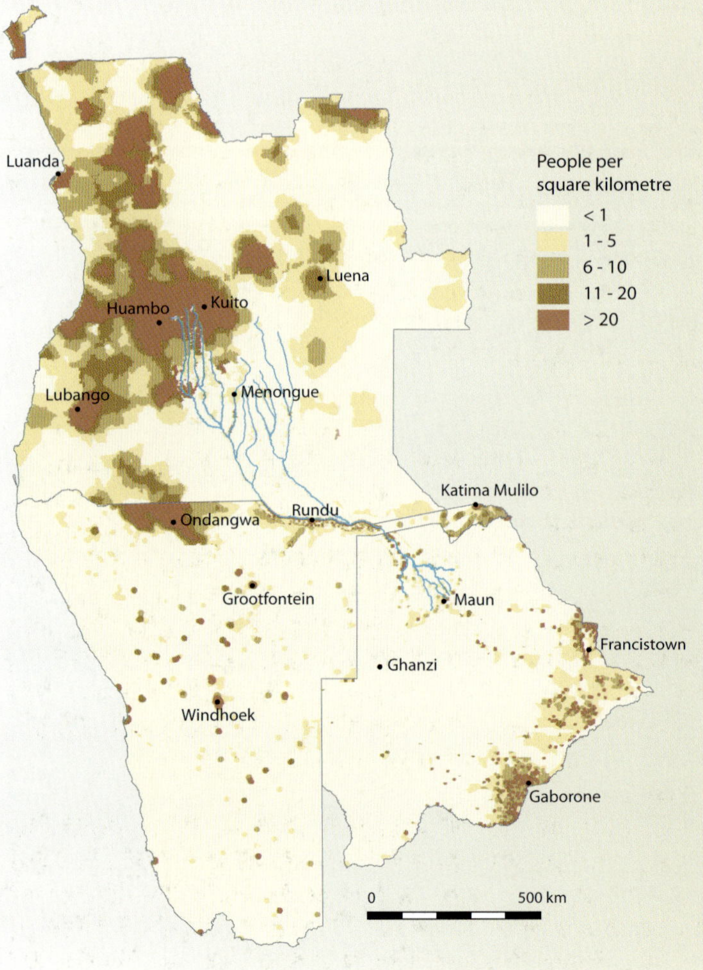

River water in both sub-basins is generally crystal clear. Very few suspended particles and dissolved substances – including nutrients – are carried in the water. Much of the purity is due to the substrate of Kalahari sand, which consists mainly of grains of quartz. Little can be dissolved from the inert quartz and the sands have been leached by relatively abundant rain over millions of years. Very few clay particles are therefore available to be washed into the river water. The paucity of suspended clay in the flow off the catchment is fundamental in maintaining the permeability of the Delta's substrate and its fresh water (*see page* 55).[4]

But there is another, less obvious reason for the clear, nutrient-deficient waters. This is the effect of the same vegetation that slows the speed of water. Many of the nutrients and suspended clay particles that do get into the water are filtered out by the plants. The floodplains are especially broad in the Cuito sub-catchment, but lots of the valleys in the Cubango sub-catchment are likewise congested with dense beds of reeds, as shown in the photographs on *page 30*.

Rainfall averages over 1,300 millimetres per year in parts of the catchment furthest to the north-west. From there it steadily declines as the river moves southward, until it averages 450 millimetres in the lowest reaches of the Delta **(Figure 10)**. Almost all the rain falls during the summer months between October and April, with the highest falls in December and January in most areas. However, in the most northerly areas, rain often falls in two peaks, one in November-December and the other in February-March. As a result, there are often two peak periods of flow into the Delta (*see page 41*).

Average rainfall in the middle reaches ranges between 500 and 600 millimetres per year. This zone also consists of Kalahari sand into which rain readily percolates and so there is almost no run-off, even after very heavy thunderstorms. The Okavango River thus functions as a conduit in the middle reaches, channeling water from its upstream Angolan sources to Botswana and the Delta.

The river is very much a lifeline for many people living along this channel in the middle reaches. It

Dense, fairly tall Brachystegia or miombo woodlands cover gentle hills of red Kalahari sand in the upper catchment of the Cuito River and its major tributaries. Very few people live in this broad expanse of Angola.

waters their livestock and vegetable gardens and provides fish, reeds for houses and pastures for cattle. No other permanent surface water is available in the area, and so the value of the river for local residents is particularly high. This dependency on waters of the Okavango is quite different from conditions in the upper catchment where abundant rain means that people are less reliant on water from the thousands of tiny streams that feed the Okavango. Evaporation rates are also lowest in the upper catchment, and rain is therefore more effective in supplying the moisture needs of crops and pastures than downstream. Indeed, aridity further south is as much a consequence of very high rates of evaporation as it is of low rainfall. For example, evaporation per year

While most households in both the upper catchment and the middle reaches practise some kind of agriculture, a great many families do not depend on agriculture as a main source of income. Instead, additional incomes from wages, small businesses, pensions and remittances provide for most cash and food needs. The apparent paradox of rural, farming households relying on non-farming income is a consequence of two fundamental features throughout much of the Okavango River Basin.

The first is that conditions are poorly suited to agriculture: soils have low fertility; rainfall varies too much to produce good yields from such rain-fed crops

Livestock diseases and crop pests also limit production, and so most families battle to harvest enough to meet their need for staples, let alone surpluses that might be sold to provide for their cash needs.[5]

Secondly, the scope for commercial production is limited by much of the Okavango River Basin being far from markets where occasional produce might be sold. Most people in all three countries are concentrated elsewhere in Angola, Namibia and Botswana, either in places better suited to farming or, increasingly, in towns and cities **(Figure 12)**, and *see page 71* in Chapter 7. For example, less than 3% of the total Angolan population is in the Basin, while equivalent proportions for Namibia and Botswana are also small, respectively 7% and 5%.

Nevertheless, the highest densities in the northern parts of the Basin are indeed a consequence of the relatively good farming conditions there, at least compared to areas further south. Densities elsewhere in the Basin are generally low and there are large areas where there are no, or just a handful of people. Menongue, Rundu and Maun are the only major urban areas in the whole Basin. Rundu and Maun each have populations of over 50,000, while Menongue probably has between 30,000 and 40,000 people. All other towns and villages are very small, most having between several hundred and a few thousand inhabitants.

The comparatively high population densities on the southern banks of the Okavango River in Namibia are a consequence of unusual circumstances. Perhaps more than half of this population was born in Angola or was born in Namibia of immigrants from Angola who moved south to escape the Angolan civil war, particularly during the 1980s and mid 1990s. Many Angolans were also attracted by the relatively better economic opportunities and social services in Namibia.

A consequence of the comparatively small proportions of the Angolan, Namibian and Botswana populations is that each government pays less attention to services and development in the Okavango Basin than elsewhere. The centres of government in each country – Luanda, Windhoek and Gaborone – are also all remote from the Basin and its needs. With the notable exception of tourism to the Delta (*see page 81*), enterprises and resources that generate significant revenue are also far from the Basin: for example, services in Luanda, offshore oil, and diamonds in the Lunda provinces in

Most homes are clustered in villages or small towns in the Angolan headwaters (*opposite page top*) and downstream around the Delta (*above*), whereas households are generally spread as solitary homes along the river valley in the middle reaches (*opposite page bottom*).

services in Windhoek and Walvis Bay, and tourism in western Namibia. As for Botswana, services are concentrated in and around Gaborone, while exports of diamonds and beef are mainly from the central and southern regions.[6]

These factors – small populations, remoteness and other economic interests – mean that the Basin enjoys less interest than it should. For Angola, much of the catchment remains inaccessible as the *terras do fim do mundo*. Plans for hydroelectric and large-scale irrigation schemes have often been suggested, but none has materialized. There are no dams of any significance on any of the Okavango's tributaries in Angola.

There are also no dams in Namibia. Despite the Okavango's poor soils and isolation from markets, the Namibian government's main interest in the river is for irrigation. This use is often justified with the assumption that Okavango water can be used to create Namibia's breadbasket. Although there have been dozens of agricultural development projects in recent decades, farming remains unproductive and irrigation projects

Fires often burn over south-eastern Angola, as shown so clearly in the photograph on the cover. Tall savanna woodlands are being reduced to shrubland in many areas, both by frequent, intense fires and by uncontrolled logging.

to the central regions of the country and to use the water for hydropower, but these have not been implemented. In summary, the Namibia government mainly sees the river water as a passing resource to be harvested before it is lost at Mohembo.[7]

Although some areas were proclaimed as hunting reserves by the Portuguese administration, no land within the Angolan catchment is now managed for environmental conservation. One small conservancy and the Bwabwata National Park, which includes a stretch of about 20 kilometres of the Okavango River, are the only conservation areas in Namibia. In Botswana, by contrast, most land in and around the Delta is under conservation management, either in the Moremi Game Reserve or as wildlife management areas (*see page 79*). Likewise, there is almost no tourism in Angola, significantly more in Namibia, and then a substantial tourism industry in Botswana (*see page 81*).

Almost all land in the Basin is formally owned by the state, and except for national parks, government and private farms, formal urban areas and concessions areas, most land is settled and used as communal land. Access to this land is largely controlled by traditional authorities in Angola and Namibia where neither the state nor traditional authorities place effective limits on the use of commonage resources. In Botswana, access to land is controlled by land boards, and the use of commonage resources is controlled to some extent in designated wildlife management areas. Nevertheless, it is generally in everyone's interests to exploit commonages maximally throughout the Basin (*see page 79*). Wealthy residents who have lucrative off-farm incomes graze as many animals as they can manage, often

at the expense of poorer folk who subsist entirely on farming. A lack of control over commonages also leads to excessive clearing of virgin land (some of which is never used for cropping), run-away fires, and logging.

Recent history has treated residents of the Basin in the three countries quite differently. Angola recently emerged from a long period of strife that began with its war of independence from 1961 to 1975 which was then followed by a civil war from 1976 to 2002. And before those 41 years of turmoil, Angolans suffered from centuries of forced labour and slave trading. Namibia has had an easier, calmer past, although it too endured several decades of armed struggle for liberation from South African administration, which ended with independence in 1990. Botswana has had the most tranquil history, being the first to gain its independence in 1966 and having never had a war. Its government has been stable, and has enjoyed a reputation for service delivery and respect for human rights.

People in the upper catchment have thus endured the most violence, the highest levels of insecurity and the poorest public services. Moving downstream, history has been kinder in Namibia, and even gentler in Botswana. Together with these differences, the value of water increases downstream from Angola to Namibia and Botswana as the Okavango River enters environments that are progressively drier. Likewise, the value of tourism and wildlife also increases downstream because each of most attractive as an oasis downstream, and nutrient resources that support wildlife are most plentiful in the Delta. The downstream Delta is the part of the Basin managed most effectively for conservation, but it is at the mercy of activities in the upstream Basin where incentives for environmental management have yet to develop. Creating incentives to manage the whole Basin wisely is one of the greatest challenges for people who value the Delta.

KEY POINTS

1. At the terminus of the Okavango River Basin, the Delta depends on the healthy flow of water from its catchment in Angola and the safe passage of water through Namibia. What is called the Okavango is therefore much more than the Delta.
2. About half of the Delta's inflow comes from the Cubango sub-catchment and the other half from the Cuito and its tributaries in Angola. Flows down the Cubango are quite rapid, whereas slower run-offs down the Cuito reach the Delta later.
3. Water flow is slowed and filtered by large expanses of reeds and floodplain grasslands that line the major tributaries and rivers in Angola.
4. Although rural households farm with crops and livestock, off-farm incomes from wages, business earnings, remittances and pensions usually provide for most of their livelihood needs.
5. Small proportions of all people in Angola, Namibia and Botswana live in the Okavango River Basin, and each country's greatest economic and political interests lie outside the Basin.
6. Although agricultural and hydro-power developments are often proposed in the Angolan catchment, much of it remains undeveloped and prisitine. Namibia often sees the Okavango's waters as a transient resource to be harvested before entering the Delta.
7. Waters of the Okavango gain increasing value as they flow downstream. History has treated residents of the Basin more kindly downstream, where the waters have greatest

4

The functioning Delta

Some people compare the Delta to an organism. Water enters the Delta through its mouth and oesophagus, the Panhandle. Further down, arterial channels distribute water across the alluvial fan.

Island plants work as kidneys to remove salts. Sediments accumulate as plaque in the channels, later creating blockages equivalent to clots in the arteries. When water is forced to bypass the blockages, new channels are formed which feed water to places that have not recently been flooded. And like body fat reserves, nutrient resources have accumulated in these places over very long periods.[1]

This chapter explores three of these metaphorical systems in some detail. The first is the blood system, to develop an understanding of how water is distributed across the Delta, and how the spread of water varies in time and space. Second, are the processes that remove salts and keep the waters of the Delta fresh. Third, are the fat or nutrient reserves, to see where they come from, how they are stored and how they are mobilised to produce the wealth of life that characterises the Delta.

The flow of water

Flows into the Delta through the Panhandle are usually smallest between September and November when they are about five times lower than the highest volumes of water coming in between March and April **(Figure 13)**. In many years there is also a small, early surge in December or January which is then followed several weeks later by the main seasonal peak. The first pulse comes very largely from the Cubango sub-catchment (*see page 28*) because its flows are rapid and it often receives early rains before a short dry spell in mid-summer, while the later inflows come from both the Cubango and Cuito sub-catchments. Those from the sluggish Cuito contribute progressively more water as the year wears on, especially during winter and autumn. But overall, the two sub-catchments provide roughly equal proportions of all water flowing into the Delta **(see Figure 11**, *page 29***)**.[2]

The gauging station at Mohembo where Botswana measures the volume of water entering the Delta.

Seasonal changes in water levels, reaching up to two metres, are greatest in the Panhandle because the spread of water is confined by dense reeds and the elevated margins either side of the Panhandle. From there, fluctuations progressively decline downstream as the water spreads out across the alluvial fans. However, levels along the Kunyere and Thamalakane faults also change substantially during the year because of the damming effects of the faults (*see page 20*).

Having entered Botswana, water takes one to two months to meander through the main 190 kilometre-long channel of the Panhandle from Mohembo to Seronga. And from the head of the alluvial fan at Seronga, the water then takes another two to three months before reaching the lowest, most distal reaches of the Delta along the Thamalakane, lower Kunyere and Khwai Rivers. Flows in the Thamalakane River at Maun are thus highest in July, August and September **(Figure 14)**. The distance along the shortest path of flow from the top of the alluvial fan to the Thamalakane is 210 kilometres.

Water flow is most obvious along the many channels throughout the Delta. These are widest in the Panhandle and in the apex of the alluvial fan, whereas the channels narrow progressively downstream into the fan. They also become increasingly liable to closure and being diverted in new directions, as we shall see below. Flows in the channels are relatively rapid: usually at speeds of 120 to 300 centimetres per second in the Panhandle, 40 to 80 centimetres in the major channels of the alluvial fan, but only 10 to 20 centimetres per second in the small, distal channels.

While water is most conspicuous in the channels, there is also considerable leakage from the channels into the surrounding back swamps and

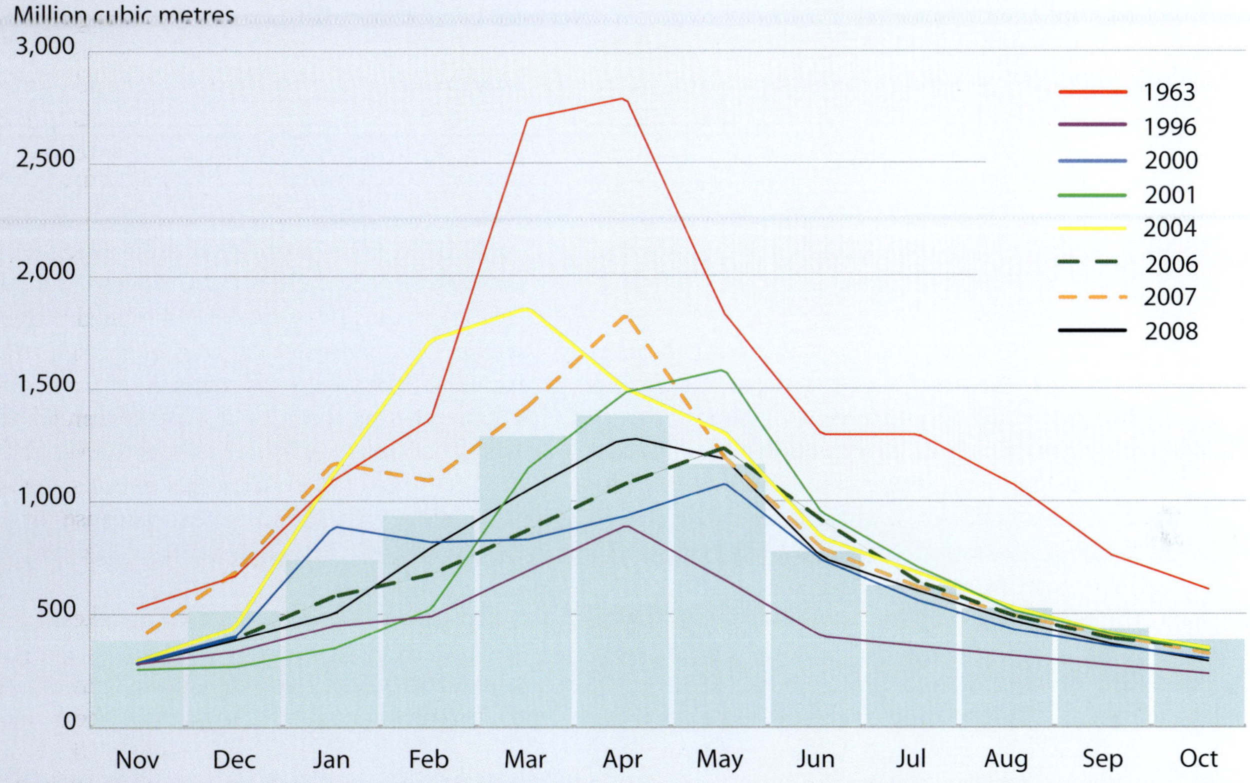

Figure 13 | *Average monthly flows of water measured at the head of the Delta at Mohembo* (blue columns), *and examples of flows each month for a number of years. Note how peak flows are early in some years and later in others, and how flows in the dry season are much more consistent than those during periods of high flow. The highest recorded inflow was in 1963 while the lowest was in 1996.*[3]

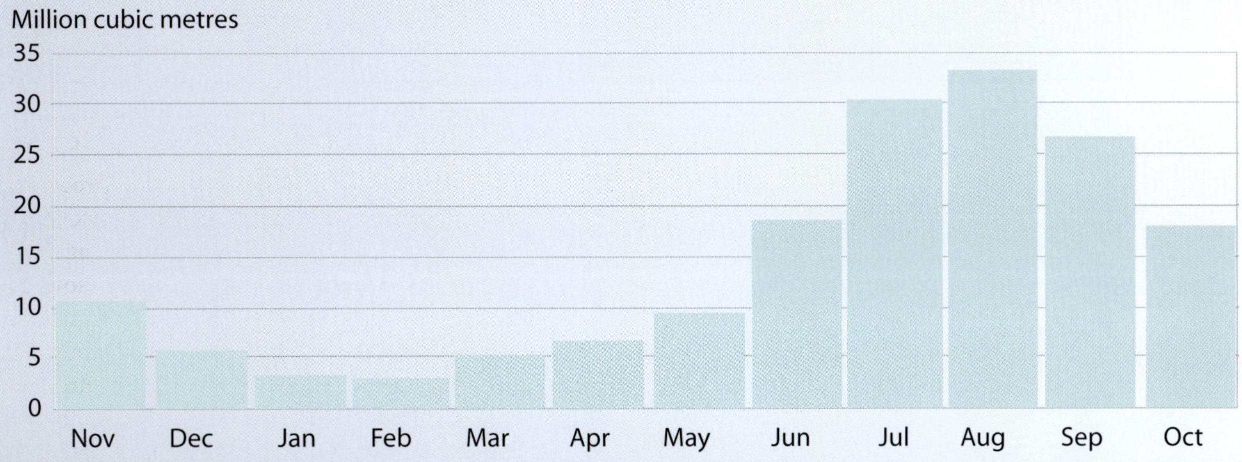

Figure 14 | *Average monthly flows at Maun, showing how levels in the Thamalakane are highest several months after the peak intakes at Mohembo. These are averages, and in many dry years no water reaches the Thamalakane. Note that the y-axis scale covers a tiny range compared to the scale for Mohembo above.*

floodplains. Some of this is through long tunnels maintained by hippos under the papyrus that lead between the channels and isolated lagoons or lakes. But most leakage is through the permeable margins of the channels from where water slowly percolates out through the extensive beds of papyrus, reeds and sedges. Here the flows are extremely sluggish – at less than 1 centimetre per second – as a result of the hindering effects of plants and shallow gradients, which become ever gentler away from the main channels.

Much of the water that filters into back-swamps remains there until it eventually evaporates, is transpired into the atmosphere by plants or seeps away into the ground. The rates of infiltration can be high, amounting to between 10 and 40 centimetres of water per day as a result of the predominance of permeable Kalahari Sand.[4] As we will see, this percolation is a critical first step towards maintaining the Delta as a freshwater system.

Annual variation in water flow and the extent of flooding

Flows into the Delta have been monitored since 1934. From these records, average intakes amount to 9.3 million cubic metres per year, which is equivalent to 9.3 cubic kilometres of water. The lowest intake of 6 cubic kilometres was in 1996 while the highest of 16.4 cubic kilometers was recorded in 1963 **(Figure 13** and **15)**.[5] This variation is entirely due to year-to-year changes in rainfall in the Angolan catchment.

Hippo tunnels under the papyrus beds aid the distribution of water across the Delta, as do the many surface paths worn by hippos, elephants and other large mammals.

Many lakes or *madiba* (singular *lediba* in Setswana) lie along belts of ancient meandering channels that were subsequently blocked or diverted. The lakes are thus remnants of deep channels, and remain as open water because they are too deep for the growth of reeds, papyrus and other emergent vegetation.[6]

The advancing spread of water over the Delta between March and October 2009.[7]

Two features stand out in **Figure 15**. One is the high degree of year-to-year change, much of which appears erratic and unpredictable. Years of high flow may be followed with equal likelihood by further good intakes or by very poor flows. The second obvious pattern is the cycles. While these are made clear by the running average line, it is not that easy to determine their length or periodicity by eye. However, statistical analyses of these figures from Mohembo – as well as from rainfall at Maun and in western Zambia, water flows along the Zambezi River, and signatures of climate in stalagmites, pollen and tree rings – indicate that rainfall and river flows follow two cycles: one lasting from 16–18 years and another 60–80 years.[8] It seems that a new period of higher rainfall and greater inflow started during the mid-1990s, but whether this will continue over the next two or three decades remains to be seen.

Volumes of inflow largely determine the extent of flooding or inundation, so the spread of water is limited in years with low inflows, while the biggest and lasting flooding occurs when flows are highest. Examples from the interpretation of satellite images in **Figure 16** show how the extent of flooding varies. Data obtained from those kinds of analyses and extrapolations from water inflows indicate that at the height of flooding an average of about 9,000 square kilometres is inundated. However, in 1996, which was the driest year on record, a maximum of about 5,300 square kilometres was under water, while about 15,500 square kilometres was flooded in 1963.[9] These are the extents of flooding at its maximum each year. Areas of inundation range between about 5,000 and 6,000 square kilometres during low water months before the annual pulse of inflow.'

In addition to inflow, the expanse of flooding in any one year is influenced by two other factors. One is the extent of inundation in the previous year, since flood waters spread further on top of existing sheets of water or if the soil is already relatively saturated. Rainfall over the Delta also affects flooding, both by adding to surface inflow from the catchment and by wetting the soil so that less new floodwater infiltrates the sands. Separating out these three influences on the extent of inundation shows that rainfall accounts for 21% of the variation in flooding each year, the extent of previous inundation explains another 1%, while inflows account for the major share of variation: 49%.[10]

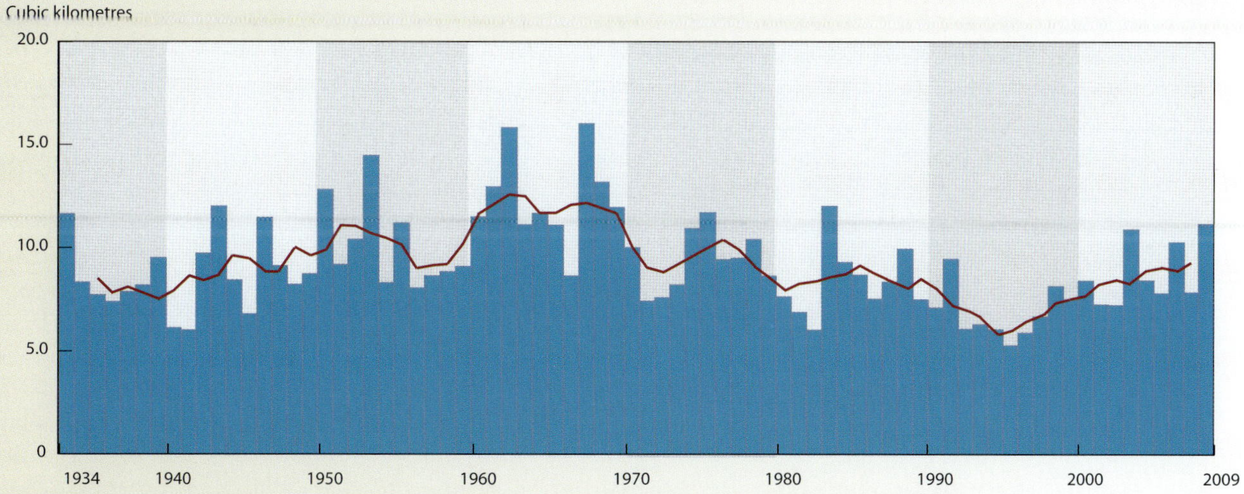

Figure 15 | *The total volume of water entering the Delta each year at Mohembo. The line is a 5-year moving average to help highlight trends and cycles. Note: the flow given for each year is actually the total flow between November of the previous year and October of the given year; in other words from the beginning to the end of each season of inflow.*

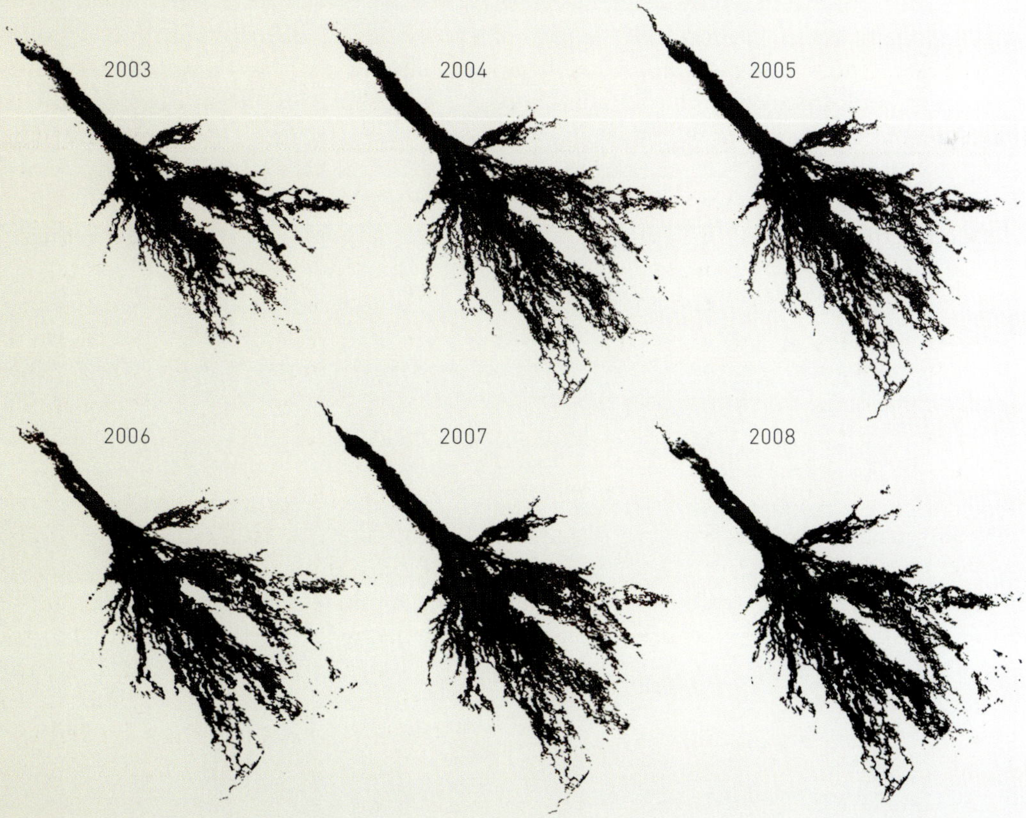

Figure 16 | *The maximum extent of flooding each year from 2003 to 2008. Most of the annual variation in the extent of flooding is in the southern and western areas. The eastern areas are flooded more permanently and regularly, by contrast.*[11]

Rainfall in the northern parts of the Delta averages about 550 millimetres per year, whereas Maun and Sehitwa in the south receive about 450 millimetres. Most of this falls during afternoon thunderstorms between November and March **(Figure 17)**. Since this is several months before the annual floodwaters spread across the area, animals and plants in the Delta enjoy an unusually long season of at least some wetness. By contrast, rain-fed wetlands elsewhere in southern Africa only have water for a few months. Biological production in those wetlands is thus limited to much shorter periods than in the Delta.

Rain is of particular importance as the main source of water for the growth of dryland vegetation scattered in and around Delta. Thus, the growth of grass and many other plants depends on precipitation. Numerous pans and pools in these areas are only ever filled by rain, and they often hold the only available fresh water for many miles around. The length of time the pools hold water depends largely on the amount and frequency of rainfall. Those that fill after sporadic heavy falls during spring usually dry up quickly if further rain does not fall, since evaporation rates are also highest in early summer **(Figure 17)**.

Of all the water entering the Delta, an average of only 2% ever leaves as surface flow. Much of this is down the Boteti River, but significant losses can occur along the Kunyere River into Lake Ngami and Khwai River into the southern areas of the Mababe Depression. These outflows only happen during good years when flood waters reach that far south. Similarly, only about 0.2% of water ever sinks into the deeper groundwater. Of the remaining 98%, about 74% evaporates directly from surface waters, with the highest rates of evaporation being in spring and early summer **(Figure 17)**. The other 24% is also lost to the atmosphere, but only after first percolating into shallow ground waters and then later being transpired by island plants, as discussed below.[12] It is the predominance of permeable Kalahari Sand that allows so much water to percolate and be available for the growth and transpiration of island vegetation.

Most of the water flow down the Panhandle is along a single, main channel. Upon reaching the alluvial fan, the channel splits into four main distributaries, from east to west, the Selinda, Nqoga, Jao and Thaoge channels. These channels then split into further distributaries downstream **(Figure 18)**. However, flows are not divided equally between the four main channels, and

Figure 17 | *Average monthly rainfall and evaporation at Maun. While rainfall varies greatly, evaporation rates are rather similar from year to year. The overall aridity of the environment around the Delta is more an effect of high evaporation than low rainfall. Thus, total evaporation at Maun – at 1,884 millimeters per year, on average – is almost four times the average rainfall of 448 millimetres.*

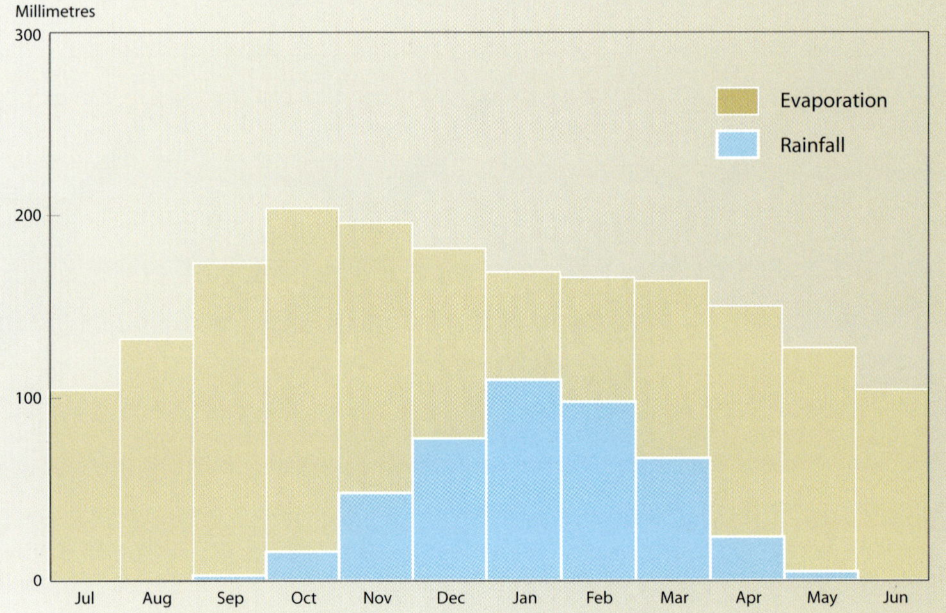

Most of the Delta's rainfall buckets down during dramatic afternoon thunderstorms.

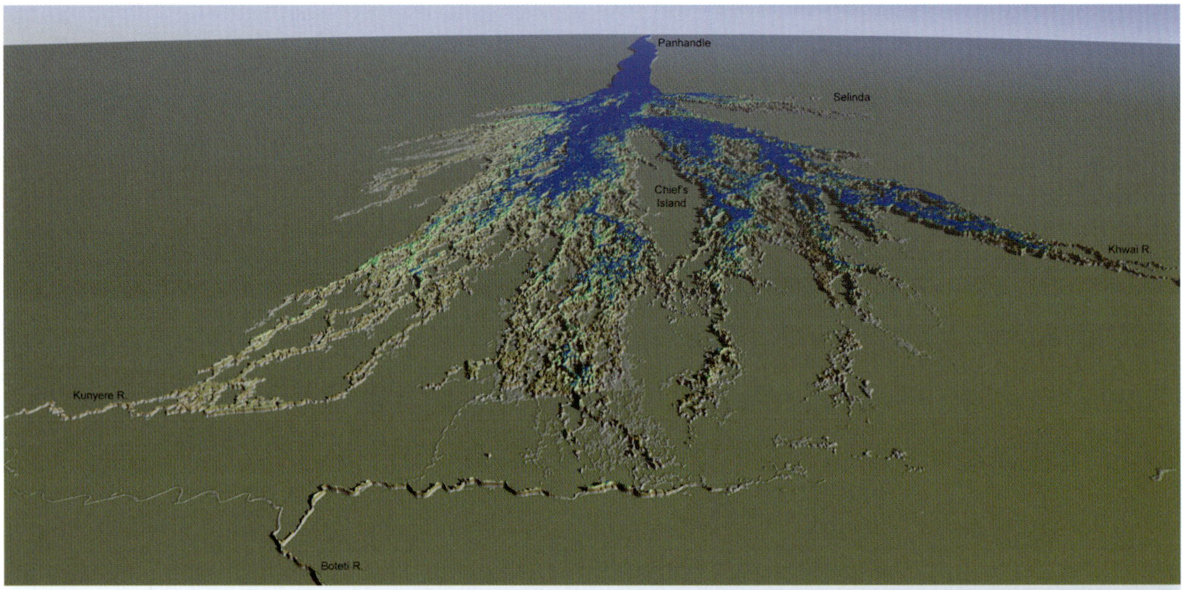

Figure 18 | *A model of flooding frequency based on an analysis of satellite images taken between 1989 and 2007.*[13] *Places that were inundated most often are higher and bluer than those that seldom received water.*

therefore do not radiate out equally across the alluvial fan. In recent decades, most water has flowed to the south-east into the Nqoga, Mboroga, Manuchira and Khwai areas north of Chief's Island, and also into the central Jao and upper Boro. By contrast, flows north-eastward to the Selinda channel and south into the Thaoge and Karongana areas have usually been weak, and have only occurred in years with exceptional floods.

There are two reasons for the unequal distribution of water from the head of the fan. One is the result of blockages, especially along the Thaoge (**Figure 19**, and see the account below of how these obstacles redistribute water). Another factor is the current location of the main Panhandle channel close to the eastern ridge that separates the permanent swamps from the higher adjoining woodlands. As a result, water percolating eastwards out of the main channel soon dams up because it can't spread beyond the ridge. However, there are no immediate barriers on the other side where large volumes of water can filter out of the main channel to make their way slowly west and then south. What this means is that flows to the west and south first have to permeate through extensive reed beds before reaching the southern extremities of the Delta. By contrast, flows to the south-east and east are more channeled and regular, and therefore sustain large areas of permanent swamp north and south of Chief's Island.

The changing distribution of water

The broad zones shown in **Figure 19** reflect flooding conditions seen in recent decades. While these conditions change seasonally, the changes are usually rather predictable as flows from Angola rise and fall. But events and processes occurring within the Delta also alter the distribution of water, but they do so in ways and directions that are largely unpredictable! These unexpected changes result in wetting and drying that are markedly different from those experienced normally. Expanses of swamp may suddenly dry up when water diverts elsewhere, for example, and new floodplains may develop in places that have been arid for decades.

Most distributional changes are due to blockages of channels which forces water to take new directions. Sites where major blockages are known are shown in **Figure 19**, but those on the

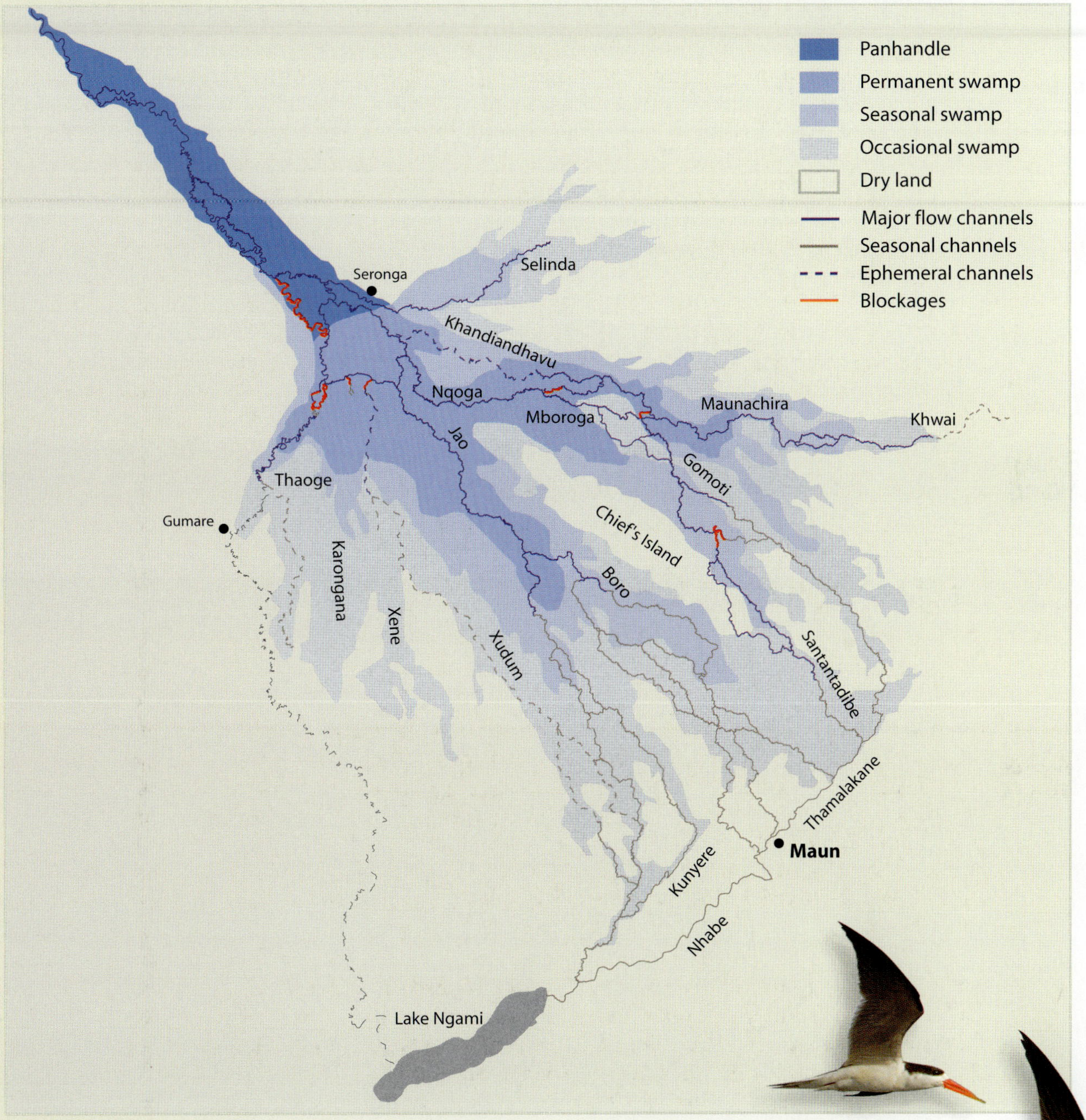

Figure 19 | *The main channels and flood zones of the Delta. As the name implies, the Permanent Swamps hardly vary in size from year to year and usually cover about 3,000 square kilometres. Annual changes in the extent of flooding are thus largely in the Seasonal Swamps and Occasional Floodplains. Boundaries between these two zones are usually hard to define, and they also change during wet and dry phases. For instance, the Seasonal Swamps become much more extensive during repeated years of high inflow.*[14] *Vegetation growth and types are closely related to levels and frequencies of inundation, and so the categories of plant communities described in Chapter 5 are closely linked to these zones.*

Thaoge provide the most impressive example of what happens once a channel stops flowing. Up until 130 years ago, much of the water entering the alluvial fan flowed down the Thaoge and its distributaries, thus regularly flooding what are now the dry grasslands of the western half of the Delta. Flows down the Thaoge also reached and filled Lake Ngami. Indeed, Charles Andersson boated up the Thaoge from Lake Ngami for 13 days in August 1854, travelling an average speed of eight kilometres per day.[15]

Then in about 1883 and 1884, the Thaoge became progressively blocked by accumulations of floating vegetation – or sudd – and encroaching vegetation, especially papyrus. And despite numerous attempts to open and canalise the channel it still remains largely dormant.

Probably as a result of flows along the Thaoge being halted, flooding to the east then increased, raising water levels in the Maunachira, Mboroga and Santantadibe. However, the northern reaches of the Mboroga became blocked in the 1970s, which then led to reduced flows into the Santantadibe and Gomoti, but also greater inputs to the Khandiandhavu, Maunachira and Khwai distributary areas.

At least some of Lake Ngami's former glory has now been restored through a new feeder. In about 1996, flows into the Xudum and Xene began and have continued to increase until the large inflow from Angola in 2004 eventually pushed water down the Kunyere River and into the Lake **(Figure 20)**.[16] Just what caused increased flows into the Xudum and Xene is

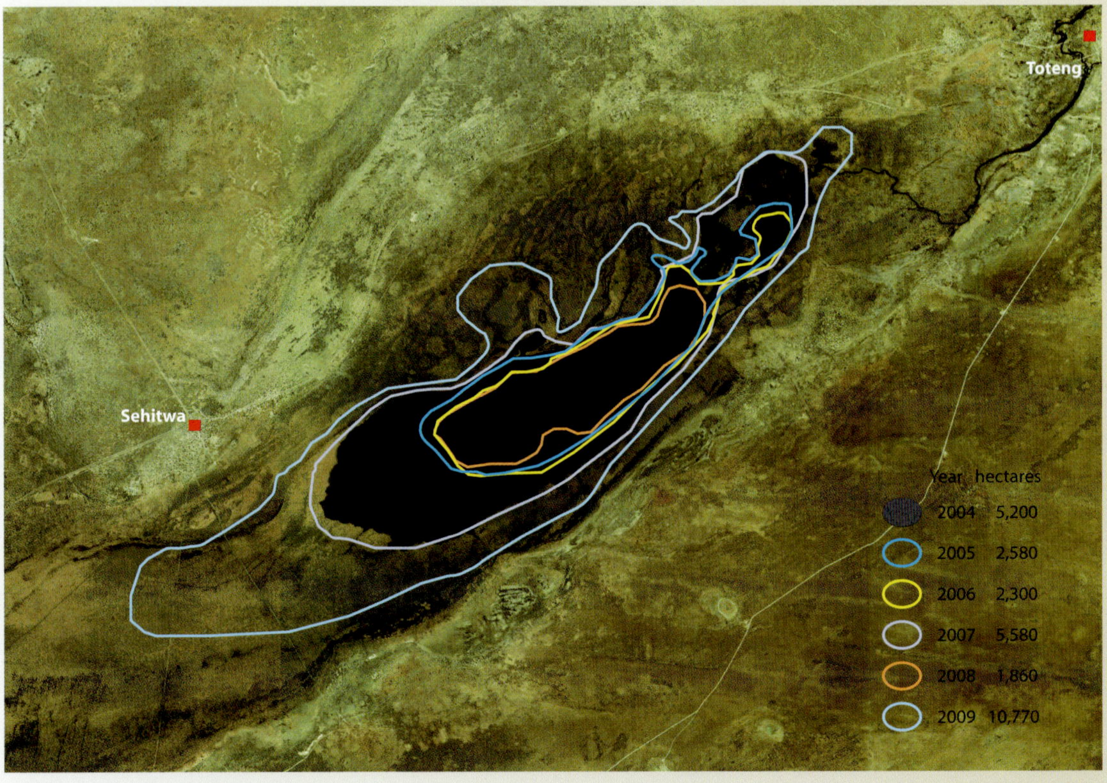

Figure 20 | *The black expanse is water in an image of Lake Ngami taken in September 2004, which was the first year that the lake had had water since 1989. Levels of the lake were lower in 2005, 2006 and 2008, rather higher in 2007 and then much higher in 2009, as shown by the lines marking the expanse of water in each year.*[17]

unknown, but many people speculated that it was due to a small tectonic event. This is quite likely in such a flat landscape, where a shift of a few centimeters up or down could cause flows to speed up or slow down considerably. Nevertheless, firm evidence has yet to be found for tectonic movements causing recent distributional changes anywhere in the Delta.

The best explanation for changes in flow directions starts with, and rests on the fact that almost all the solid, sandy sediments carried in from Angola remain in, and close to the beds of channels. Here, the solid, mostly quartz grains are gently rolled along as far as the speed of water permits. However, speeds slow as increasing proportions of water leak out of the channels, and this causes more and more sediment to be deposited and the channel beds to rise in a process called aggradation (see **Figure 21**). Dense margins of papyrus stabilize the position of the rising channels, thus preventing them from shifting to lower adjacent ground. And as the water slows even further, papyrus growth impinges on the channels and floating sudd and debris increasingly get stuck in the channels.

As blockages become more and more impermeable, water in the channel is dammed and starts to rise. At some point, it breaks through the walls of the channel upstream of the obstructions, and makes its way along newfound courses into the surrounding, lower-lying floodplains. This is known as channel avulsion. It is widely held that these new courses often follow the paths worn through the swamp vegetation by hippos. However, fire almost certainly plays a major role in charting new directions of flow by burning away elevated levees of dry papyrus. Large channels may thus be diverted into new waterways which go off in quite different directions.[18]

Other factors thought to cause flows to change direction include the subsidence of the ground, perhaps as a result of tectonic shifts, and simple damming due to the build-up of floating sudd and encroachment of aquatic plants without the aggrading effects of sediments.

Changes in the distribution of water often cause annoyance, especially when water no longer reaches places that have enjoyed consistent flooding and flows. Those affected, for example villages that lose fishing grounds or lodges that no longer have access to channels, observe the losses as permanent and sometimes blame other people who use the Delta.

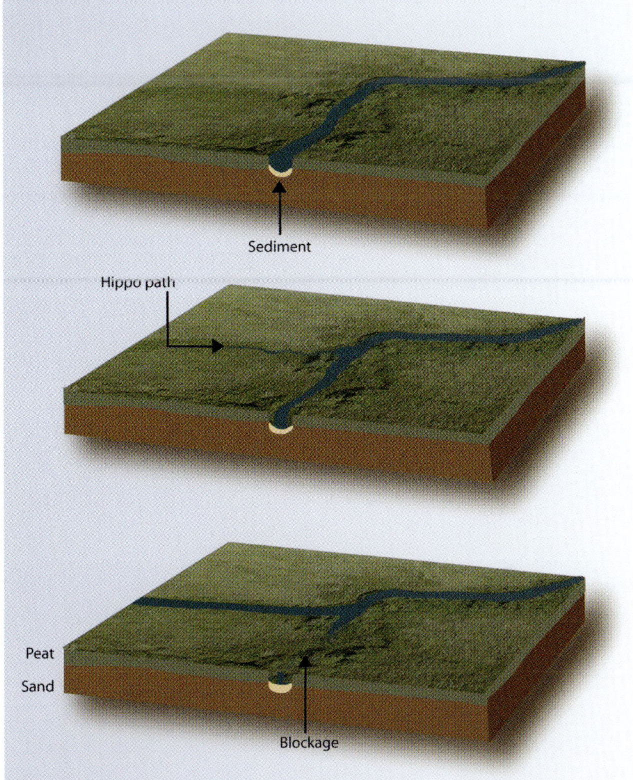

Figure 21 | *Snapshots during the development of a blockage along a channel and the subsequent diversion of its flow along a newly formed channel. Note the rising levels of sediment and the narrowing of the main channel, and the development of the blockage as the speed of flow decreases.*

However, such changes are normal and temporary in the longer term.[19] The changes are also inevitable in this flat, aggrading landscape and they contribute to the greater productivity of the Delta. Indeed, significant features of the Delta are consequences of the distributional changes:

- Habitat diversity is increased
- Habitats change rapidly
- As a result of drying, nutrients trapped in aquatic plants are released for use by other organisms
- When inundated, nutrients lying dormant in sediments are mobilised into new cycles of production (*see page 62*)
- Once flooded, mounds built by termites become islands on which transpiration by trees helps to concentrate salts and thus maintain the fresh waters of the Delta.

The blockage of water flow and its redistribution to other areas of the Delta is crucial to the functioning of the Delta. Sudd (*2nd from bottom right*), encroaching papyrus (*bottom right*) and floating debris (*2nd from top right*) all help clog channels, which later become overgrown by papyrus and other aquatic plants (*top right and bottom left*). The channels are also raised as a result of sediment deposition so that they later stand above the surrounding floodplains (*top left*).

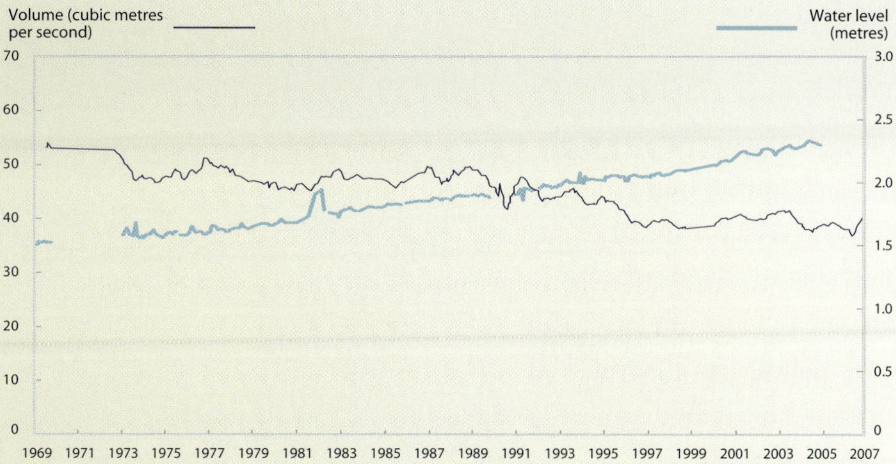

Figure 22 | *The gradual build-up of the Nqoga channel at Gaenga between 1969 and 2007. Rates of sand accumulation can be substantial, perhaps as much as five centimetres per year, or half a metre in 10 years. Water levels then rise* (right axis) *causing the volumes of water discharge to decrease* (left axis).[20]

The quality of water

Anyone visiting the Delta or using its water must be astonished at its clarity and purity, qualities that stem from the nature of water coming off the catchment (*see page 32*). Here, few particles or nutrients are available to be washed into the Okavango River and, in any case, dense barriers of aquatic plants trap much of any material that does get into the river. To put the purity in context, the conductivity of water entering Botswana is 15–20 times lower (i.e. better) than would be classified as *Ideal* water for drinking, and about 40 times better than water of *Acceptable* quality, as established by standards of the Botswana government.

Conductivity is a measure of the quantities of salts and other solids dissolved in water. Of course, there are other dimensions to water quality, such as its colour, odour, turbidity and acidity but, again, all the monitoring work done so far indicates that the inflowing water is exceptionally clear of dissolved and suspended materials. Average pHs of between 5.9 and 7.6, or near neutral, are measured throughout the Delta. Slight acidity or alkalinity is sometimes detected locally and is usually the result of the decomposition of organic matter or of concentrations of algal growth, respectively.

These comments refer to concentrations of chemicals and particles, and not to the quantities of material swept into the Delta. These are indeed substantial when the tiny concentrations in the huge volumes of water are added up. In fact, an estimated 170,000 tonnes of bed load (mainly minute grains of quartz sand), 30,000 tonnes of suspended material (largely clay and organic matter) and some 381,000 tonnes of salts are carried down the Okavango River each year.[21] What happens to all these tonnes?

Because water speeds are so slow in its meandering channels, about 90% of the sandy bed load settles in the Panhandle, while most of the remaining 10% is deposited close to the head of the alluvial fan. Much of the lighter matter in suspension also remains in the permanent swamps of the Panhandle and at the top of the alluvial fan. That entrapment and the fact that so few clayey particles come down the Okavango River is crucial in retaining the permeability of Delta's substrate.[22] Lots of surface water can therefore sink into the groundwater which then allows for the concentration of salts beneath islands, as described below. For example, rates of infiltration in one seasonal floodplain ranged between 45 and 54 millimetres of water per day, which was 9 to 10 times higher than the amounts of water lost by evaporation.[23]

Most dissolved salts eventually permeate through the sandy substrate and into the groundwater. This happens after they have been distributed by surface waters, and their concentrations increase progressively as a

Water almost everywhere in the Delta is so clear and clean of salts, nutrients and floating particles that it surpasses the quality of bottled water.

result of surface evaporation. Concentrations in samples steadily increase by an average factor of 10 between the head of the Panhandle and the distal reaches of the alluvial fan in the south-east. But in terms of conductivity, even the highest concentrations found at the bottom of the Delta are similar to those found in most bottled water.

Quantities of nutrients in the inflowing water are likewise very low. Most nitrogen, phosphorous, sodium and potassium, for example, is quickly absorbed by papyrus and phragmites reeds growing in the Panhandle and along the main channels at the base of the alluvial fan. Very small amounts of these elements therefore make their way to the seasonal and occasional floodplains. Here, most biological production is fed by nutrients that accumulated in their sediments over long periods (*see page 62*).

While the waters of the Okavango are usually clean and clear, local conditions of impurity sometimes develop. For example, high concentrations of salts may develop as a result of evaporation in pools that are isolated from the main Delta flows. Water percolating through the papyrus beds in the Panhandle is sometimes so depleted of oxygen as a result of aerobic respiration by bacteria, that substantial numbers of fish die (*see page 102*). And with the growing number of people living in and around the Delta, the chances of local pollution escalate, particularly in places where people are concentrated, such as at Maun and along the Panhandle (*see page 115*).

Maintaining water quality: permeable substrates, islands and sinking salt

Makgadikgadi and Etosha are open expanses of salty sediments. Why did yet another pan not form here, perhaps called the Okavango Pan? Three separate processes – all of which have to do with the removal of salts – explain the anomaly of the freshwater Delta in this arid region. The combination of these processes in one place is probably unique in the world.[24]

The first process involves the very high rates of water infiltration made possible by the limited intake of clay from the catchment and the permeability of the Delta substrate. The large volumes of water that disappear into the ground therefore carry and remove salts from the surface water.

The second process stems from the abundance of islands within the Delta on which trees and other woody plants grow. In all, there are about 150,000 islands covering a combined area of some 4,500 square kilometres. Most are tiny, consisting of just a few square metres, but others extend over hundreds of hectares; an estimated 43,000 islands are larger than 50 by 50 metres.[25] These numbers and areas provide places to grow for very many trees which draw water out of the ground and then transpire it through their leaves. As the water is pumped up through the roots, water levels beneath the islands drop, causing a substantial gradient (of 1:100 to 1:200) between groundwater levels below the surrounding floodplains and those under the islands **(Figure 23)**. The water – accompanied by its salts – thus flows laterally and continually from

...ition as water flows down through the alluvial fan. The ... the lagoon...

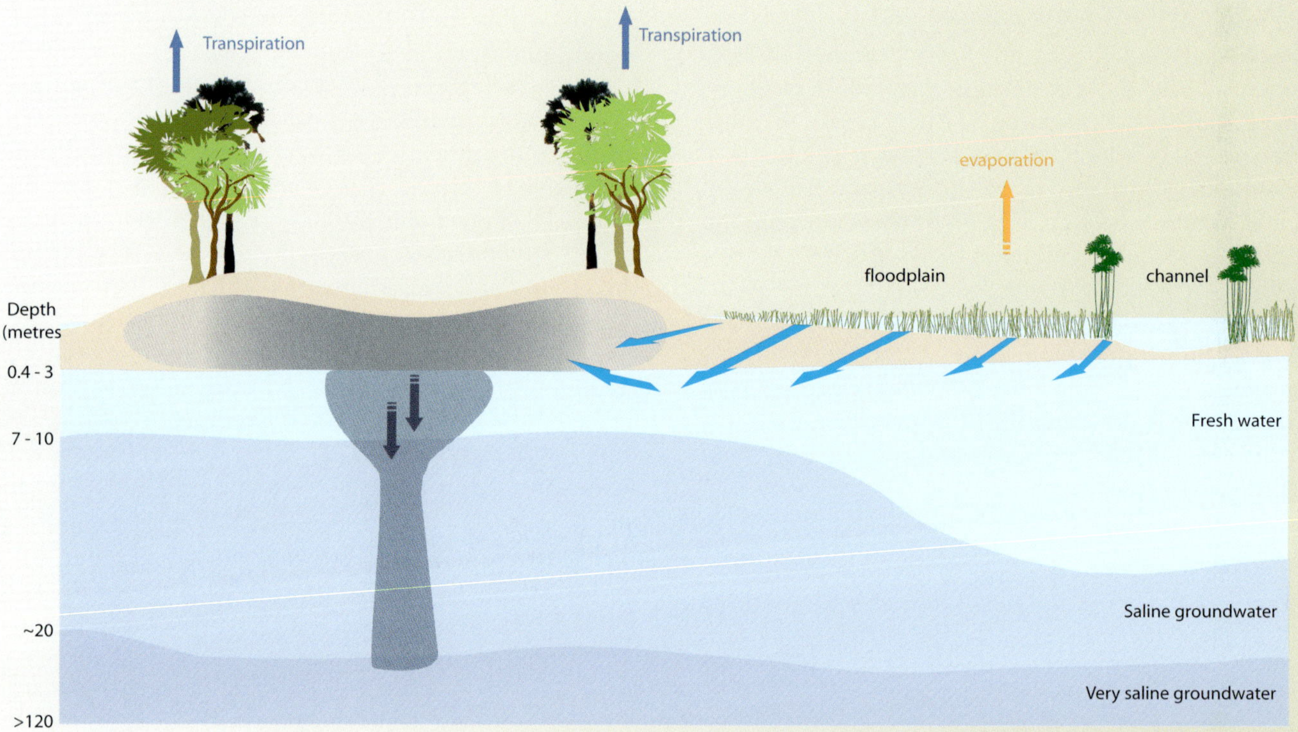

Figure 23 | *A horizontal cross-section through an island and its surrounding floodplain. Note the higher margins of the islands formed from deposits of calcium and magnesium carbonate and the accumulation of organic debris beneath the trees and shrubs. Few plants can tolerate the very salty soils in the island centres. It may take 200 years for an island to grow from a small copse of trees and shrubs into one looking like an atoll with a barren centre.*[26]

beneath the floodplains to below the islands. In essence, this second process moves salts from the floodplains and concentrates them below island vegetation.

The rates of concentration are considerable. For example, chlorine in water below the islands is 500–1,000 times more concentrated than floodplain surface water. According to their chemical properties, different salts and elements react differently once they are concentrated. Magnesium and calcium are precipitated into solid calcrete carbonates, while precipitates of silica and potassium form clayey soils that are between 3 and 10 metres deep below the islands. Dissolved organic carbon and sodium bicarbonate remains in solution in the island groundwater, but as their concentrations increase, sodium bicarbonate crusts may form on the surface and the island centres become too saline for most plants. As a result, most trees only grow on their margins, leaving the central areas of islands to grasses and a few palm trees that tolerate the salty soils.

Most islands are thought to originate as termite mounds on which copses of trees grow when the mounds are later flooded. The islands then expand and continue growing as more and more clayey deposits and calcretes are added to their volume. Organic matter, such as leaves, that accumulates beneath the trees further contributes to their growth. While some islands remain small and rounded, reflecting their origins as termite mounds, others have grown and merged with other islands into large, irregularly shaped areas.

The majority of the approximately 150,000 wooded islands are in the occasional floodplains where they make up about 50% of the area. By contrast, about 25% of the seasonal swamp area and only 5% of the permanent swamps are covered by islands. Most islands grew from termite mounds (*bottom left*), but others formed on ridges where old channels were raised by aggradation above the swamps. The relics of the channels are now [illegible] (*left*).

The third and final process in removing salts from the Delta also occurs below the islands. Once particularly high concentrations are reached, the saline groundwater becomes denser, and therefore heavier, than the water which lies further down. 'Density fingering' then occurs as the hyper-saline water sinks down to much deeper levels.[27] This is the final fate of dissolved salts that originated hundreds of kilometres away in Angola.

Of the estimated 381,000 tonnes of dissolved material entering the Delta each year, 56% eventually turns into precipitates that expand the islands while another 38% will later sink away as very saline water. The small remaining fraction of 6% is carried off down the Boteti River during its sporadic flows.

Nutrients

There are many aspects of the Delta that are odd. This is an oasis in a semi-desert of sand, and it is a body of water that remains fresh even though 98% of the water evaporates. Another peculiarity is that living organisms are abundant despite the fact that few nutrients enter the Delta. A measure of this wealth is provided by the number of large mammals. Based on the carrying capacity of soil and rainfall conditions at different places elsewhere in Africa, northern Botswana should be able to sustain about 1,200 kilogrammes of large mammals per square kilometre. In reality, however, the biomass of large mammals in the Delta is almost 10 times higher.[28]

Clearly, there is more to support so many animals than meets the eye. But without special equipment, eyeing the abundant nutrients that feed so many animals and plants is difficult. In addition, research on nutrients has been rather limited; while we have a pretty clear idea of the water budget, the food budget of the Delta has not been well documented.

Three nutrients are essential to all plants and animals: carbon for building bodies, nitrogen for proteins and phosphorous for the molecule ATP (adenosine tri-phosphate; also called the 'universal energy currency of life') which stores energy produced during respiration. Although other nutrients are required for metabolism and growth, it is these three elements that are needed in the most significant quantities. They are thus more routinely monitored than other nutrients, and these measures provide indications of the overall nutrient status of the Delta.[29] Since carbon is usually less constraining on production than nitrogen and phosphorous, it is these two nutrients, and particularly phosphorous, that often limit the growth and reproduction of plants and animals everywhere.

Figure 24 | *The main pathways for sources and losses of nutrients in a typical seasonal swamp or occasional floodplain.*[30]

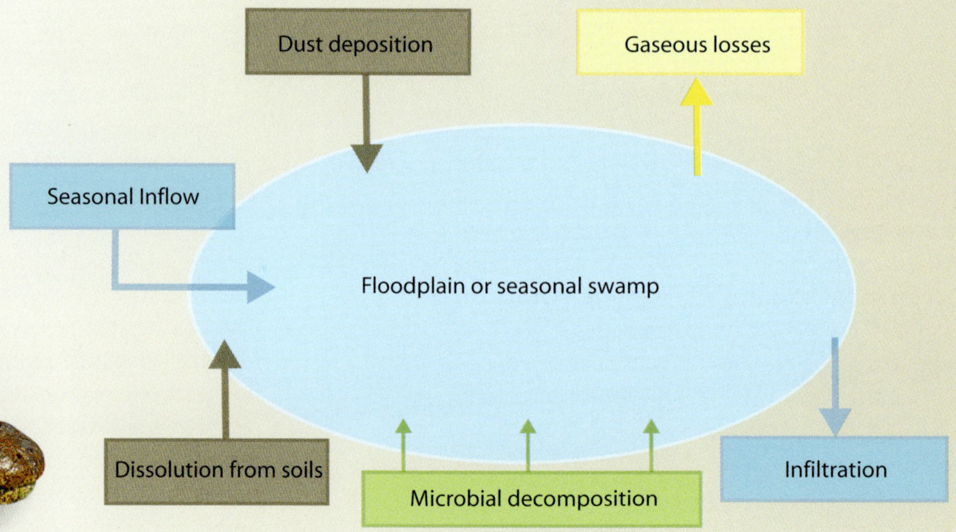

Nutrients are cycled and recycled between sediments beneath the floodplains, living swamp vegetation, peat and surrounding woodlands.

Sources and losses of nutrients

Although nutrients are limiting, how so much nitrogen, phosphorous and carbon came to be in the Delta has yet to be fully explained. However, there appear to be five main sources **(Figure 24)**. The first is the Okavango River. While nutrient concentrations in its waters are tiny, they add up to significant amounts when the volumes of water that have come to settle on the sediments over hundreds and thousands of years are considered. The concentrations of nitrogen and phosphorous measured at Mohembo are typically in the order of 1–2 and 0.0–0.1 milligrams per litre, respectively. There are no significant seasonal changes in the concentrations.

Aerosols are a second source. The Delta happens to ... anti-cyclonic conditions

that coincide with flooding. As a result, an estimated 250,000 tonnes of dust (or 570 grams per hectare) are deposited each year on the Delta.[31] Significant quantities of nutrients come with the dust, and these may contribute up to 40% of all nitrogen and 60% of phosphorous added to the Delta.[32] The aerosol nutrients settle both on water surfaces, where they may dissolve and immediately become available to biological processes, and on dry land, where they dissolve later when flooding occurs. It is almost impossible to separate out those that settle on dry land from a third supply, which is that comprised of nutrients that dissolve out of soil particles. However, this third source is probably of minor significance because inert grains of quartz comprise much of the sandy soil.

... nutrients are made available by microbes, such

Nutrients are added to the Delta's water in the form of faeces produced from food obtained in the surrounding dry woodlands.

Peat fires are not often seen but they play a crucial role in releasing nutrients that are bound up in the organic debris of which peat is largely comprised.

vegetation. They convert gaseous atmospheric nitrogen into ammonia and other forms which can be used by living organisms. Blue-green algae likewise fix nitrogen in water, and bacteria associated with the roots of the common rice grass *Oryza breviligulata* probably also fix nitrogen. Mycorrhizae of fungi found in and among roots convert ions of phosphorous into forms that can be used by plants.[33]

Finally, nutrients are imported in the faecal matter of animals that visit the Delta. This happens most often and typically by animals that drink regularly after long bouts of foraging in the surrounding woodlands and dryland pastures. Obvious examples are elephants, cattle and impala, of which there are very large numbers (*see page 106*) that dump considerable volumes of faeces and nutrients into the Delta on a daily basis. But conversely, smaller amounts of nutrients are lost when animals that have fed on aquatic food leave and defeacate elsewhere.

One of the reasons that phosphorus is so limiting is that much of it is held in insoluble compounds bound to such elements as iron, aluminum, calcium and manganese, and also to clay. Other insoluble phosphorus lies trapped in the decayed remains of microbes, animals and plants.

The relative importance of each of the nutrient sources varies across the Delta, both in time and space. For example, inflowing nutrients contribute most to production along channel margins, especially in the Panhandle. In the back-swamps, more nutrients are supplied by aerosol sources and by the limited breaking down (and thus recycling) of organic molecules bound into peat. On seasonally flooded plains, soil nutrients are the prime source once the floodwaters arrive, but deposits from dust become increasingly important as the season wears on. Production in occasional swamps probably depends largely upon nutrients lying dormant in the sediments, having accumulated there over long periods from various origins: aerosols, mineralized plant matter and faeces, for instance.

The overall abundance of nutrients means that the Delta is a sink, with more nutrients having accumulated here than have left. The losses may, nonetheless, be significant, particularly through fire which can accelerate the rate at which nitrogen is lost to the atmosphere as a gas, and through the leaching of phosphorus into deep sediments. Considerable quantities of organic carbon and phosphorous also disappear in the flow of water from floodplains to beneath islands where the nutrients remain trapped in the very saline groundwater and clays.

Dung beetles help to release nutrients originally locked into plant matter for new cycles of animal and plant production.

Nutrient cycles and uses

Most nutrient losses are, however, temporary in the sense that nutrients become locked into reservoirs where they are unavailable to new rounds of biological production. There are two principal reservoirs in the Delta: living plants, and dead peat and detritus. Nutrients locked into plants are recycled when floodwaters dry up and the plants die. This happens seasonally in areas that are inundated each year, but also in permanent swamps where large nutrient stocks are held in the expanses of papyrus, phragmites reeds and other aquatic plants. These only die off when channels change course to switch the flow of water to other places. Once the swamps dry, their degradation is often hastened by fires that burn the desiccated plants (*see page 84*). On a smaller scale, nutrients embedded in plants also are released by grazers, especially the very abundant lechwe and buffalo (*see page 106*). Dung beetles and other insects that consume manure are the first to use these nutrients, which then get cycled into other organisms that eat, parasitise or decompose the insects.

Peat in the permanent swamps holds even greater volumes of nutrients than the green vegetation above, since peat consists of dense masses of decomposed roots, rhizomes, charcoal and other organic and inorganic detritus.[34] However, nitrogen and phosphorus in peat is largely present in the form of organic compounds which cannot be readily used for growth and reproduction. Further recycling is thus needed before they are available as ammonium, nitrates and phosphates. While some of this is achieved by aerobic actinomycetes bacteria and fungi that also break down cellulose into carbon, most of the nutrient pool is only released once peat is dried or burnt.

Thus, great quantities of potential food remain inert in various forms in the Delta. The most impressive and final release of nutrients into blooms of production occurs when water reaches the seasonal and occasional floodplains.[35] The soils of the floodplains contain huge numbers of spores of anaerobic bacteria that spring to life when the waters arrive. By decomposing detritus, the living activated bacteria produce and release carbon and soluble nitrogen and phosphorous into the floodwaters.

Diatoms and other phytoplankton blossom, aquatic plants grow rapidly, and eggs that have lain dormant in the sediments hatch and develop into dense swarms of zooplankton (*see page 100*). Fish migrate into the temporary waters, spawning to produce large numbers of fish fry which feast on the zooplankton. In turn, these and the fingerlings feed countless frogs, birds and predatory fish. As the flood continues and later abates, the frogs and birds breed quickly to produce new generations, surviving young fish swim back to populate the main channels and waters of the permanent swamps, while nutrients parcelled in the bodies of insects, herbivores and other animals are redistributed elsewhere to add life to the Delta.

To conclude

It is mainly in the seasonal swamps and occasional floodplains – where the aquatic and terrestrial worlds meet – that the Delta comes to life, and its wealth is produced. The annual pulse of water mobilises nutrients which, in turn, are largely provided by the terrestrial environment. The pulsing happens regularly in the seasonal swamps and episodically in the occasional floodplains in a great number of places over a wide area. An ever-changing patchwork of different levels of nutrients, at different stages of biological succession is thus created. And all of this is aided by the extended period of water availability due to rain and flood waters being received at different times.

Wetting is thus important, but so is drying. Without desiccation, most nutrients would not be released from living plants and peat. Changes to the distribution of water make much of the drying possible and, likewise, the switching of channels allows water to inundate new ground. Most biological production occurs in places that have lain dry for some years after periods of repeated flooding in the past. The most productive sites have long histories of flooding because each flood added more nutrients that were washed in, trapped from aerosols or dropped by animals attracted to the water. When the water dries and aquatic plants die back, the nutrients lie embedded in soils and detritus, waiting – as it were – again to be mobilised into production when catalyzing water next arrives.

The wetting, drying and consumption of the Delta's larders. After the arrival of flood waters, mats of algae and emergent aquatic plants cover Lake Ngami (*left top and middle*) and water among trees near the Gomoti River (*right top*); masses of birds feeding on fish trapped in a shrinking pool, isolated as the seasonal swamps dry up (*right bottom*); and the …ding waters that lead to the death of plants and

Concentrations of greenery close to channels are due to those plants having better access to nutrients in the channels than those growing in the more sterile backwaters.

Central to the abundance of life in the Delta is this shifting kaleidoscope of water and land, with water enabling plants and animals to use larders of nutrients to grow and reproduce. Each year, floodwaters bring life but then recede so that nutrients are returned to the soil, ready for further bouts of production in the years ahead.

Because the distribution of water changes so frequently, many plants and animals have evolved opportunistic strategies. Those that are able to shift to new supplies of water do so, making the move as quickly as possible. Those that can't move, lie in wait for years, perhaps decades. And they lie waiting in a variety of forms: as eggs, seeds, spores, pupae or as aestivating, slumbering adults.

One of the best places to see and learn about episodic production is at Lake Ngami. As an outlier on the most distal margins of the Delta, the lake has often been dry. But there have also been long periods of inundation which led to the accumulation of huge nutrient reserves in its sediments. These are now augmented by the manure of cattle herds that graze around the lake, and wildlife doubtless did the same before the area was populated by stock farmers. Cattle now also graze the aquatic grasses in and around the lake, thus returning nutrients to the soil that would otherwise be trapped in vegetation. The lake waters are thus highly productive as a result of these processes that accumulate and recycle nutrients. When flooded, Lake Ngami attracts more birds than any other wetland in Botswana. For example, 5,200 white pelicans, 20,000 red-billed teal and 3,000 little egrets were counted there in 2004,[36] and the Lake supported a vibrant fisheries industry during the 1970s. The Mababe Depression is certain to be equally productive when its waters again return.

Several processes must be conserved if the Delta is to remain as productive as it now. The first is the necessity of seasonal pulses of water flow to maintain the wetting and drying that mobilize and release stocks of nutrients. Secondly, the distribution of water has to keep changing, which means that the agents that cause the redistribution of flow (papyrus growth, fine sediment accumulation and hippos) have to keep working. Third, processes that make this this an oasis of freshwater need to be sustained, for example by ensuring that inflows of suspended clays do not increase and escalating elephant populations (*see page 87*) do not decimate the riparian trees that transpire large volumes of water from islands.

These kinds of cautions come from an understanding of processes that mould the organism we call the Okavango Delta. Other cautions will be offered later in the pages ahead, and understandings of different processes will emerge as a result of further research in the years ahead. People charged with caring for the health of the Delta must be good physicians, basing their diagnoses and treatments on the best knowledge.

KEY POINTS

1. Annual pulses of water entering the Delta reach peak levels in March and April when inflows are about five times greater than the lowest discharges in October and November. The March/April pulses then take several months to flow down the Panhandle and spread across the alluvial fan.
2. Flooding thus occurs in the dry winter months after the summer rains, and life in the Delta enjoys much longer periods of water supply than other freshwater wetlands in southern Africa.
3. Inflows and the extent of flooding change substantially from year to year – and also between longer wet and dry cycles – as a result of varying rainfall in the Angolan catchment of the Okavango River.
4. The biggest floods cover about three times more of the alluvial fan than the smallest floods. The expanse of flooding is not only influenced by inflow, but also by rainfall and the levels of inundation during the previous year.
5. The confinement and deposition of sand in the channels leads to their beds rising, the water flow slowing, vegetation encroaching along their margins, and to their eventual blockage. As a result, areas of swamp dry up as the dammed water cuts new channels, often to areas that have not flooded in recent years.
6. Of the little clay carried down the Okavango

swamps. This allows the Delta's substrate of Kalahari Sand to remain permeable. The water's percolation through the sands and its transpiration by island vegetation results in the concentration of salts beneath islands. Eventually the accumulated salts sink to substantial depths below the Delta. It is these processes that keep the Delta's water fresh.
7. The Delta is a rich sink of nutrients that accumulated here over long periods from river water sources, aerosol deposits, fixation by microbes and faecal deposits of large mammals.
8. Most nutrients are however trapped in peat and living swamp vegetation. They are only *released by drying* when water is redistributed (through channel blockages and switching) and the dry organic debris is decomposed or burnt.
9. However, the nutrients are only *mobilised by wetting* into new cycles of biological production when floodwaters reach the seasonal swamps and occasional floodplains.
10. Much of the wealth of life in the Delta is therefore generated by wetting and drying caused by the repeated pulsing and redistribution of floodwaters.
11. It is vital that conservation measures maintain the processes that enable some areas to be flooded and others to be dried.

5

Plants: producers, filters and distributors

The variety of vegetation in the Delta gives it beauty, sustenance and diversity, and also controls many aspects of how this wetland works.

Water and nutrients are the primary resources that fashion the Delta. Both are inorganic, crucial to plant life and come directly or indirectly from areas outside the Delta, as described in the last chapter. Plants also require sunlight and carbon dioxide, which likewise come from remote sources. However, these two requirements are rather evenly available in any given area. The growth, abundance and structure of plants are therefore largely controlled by local variations in water and nutrient supplies. This is true throughout the world, and so it is for the Delta, as the following pages will show.

Figure 25 | *The number of plant species found in each of their preferred habitats. Plants typically found in two habitats are counted in the zones of overlap between habitats. A high proportion of plants prefer the Drylands, Riverine Woodlands, Occasional Floodplains and Rainwater Pans as habitats, while a very high percentage of aquatic species occur in both the Permanent and Seasonal Swamps. Nine more species that are not shown here all occur in Rainwater Pans, Seasonal Swamps and Permanent Swamps.*[1]

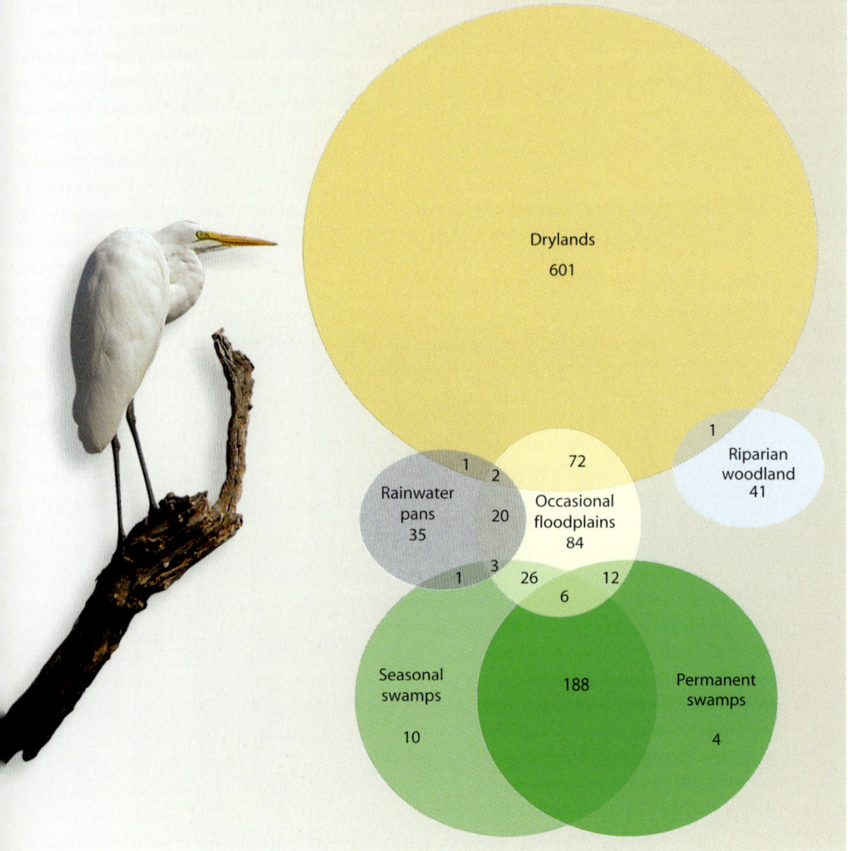

However, there is a difference! This is the important role plants perform in fashioning the living Delta, for example in causing channels to switch direction, sieving out mud and clay, and pumping water up into the atmosphere. These are local effects, which are as important as the external influences of floodwaters, nitrogen, phosphorous and other nutrients. Understanding and conserving the roles performed by plants is thus vital to the overall health and future of the Delta.

Much of this chapter focuses on higher, flowering or vascular plants which are dominated by grasses, sedges, forbs, shrubs and trees.[2] About 1,300 species have been identified so far, with the grasses (Poaceae) and sedges (Cyperaceae) being the largest groups, followed by the asters and daisies (Asteraceae) and leguminous bean and pea pulses (Fabaceae).[3]

Just over 75% of all species are herbaceous plants, while the remaining plants are woody trees, shrubs and dwarf shrubs. Eighteen species are parasites and 12 are carnivores (*see page 74*). In addition to the higher plants, there are also non-vascular plants such as mosses, algae and diatoms, blue-green algae (cyanobacteria) and dinoflagellates in this aquatic environment. Much less is known about them, and they are relatively sparse as a result of the nutrient-poor water. However, some information is available on diatoms (*see page 82*).

Do these 1,300 species constitute a particularly diverse assemblage of plants? In terms of Botswana, the answer is yes, because about ten times more species are found here than anywhere else in the country.[4] Of six major freshwater wetlands in the world, the Delta has the second highest diversity of plants, surpassed only by the Brazilian Pantanal, but greater than the Florida Everglades, Indian Sundarban, Cambodian Tonle Sap and the Australian Kakadu wetlands.[5]

Patterns of diversity

Much of the high diversity in the Delta results from wetland and dryland species growing side by side. Perhaps surprisingly, about 60% of all the plants are dryland species that grow on islands and on tongues of dry Kalahari sand that

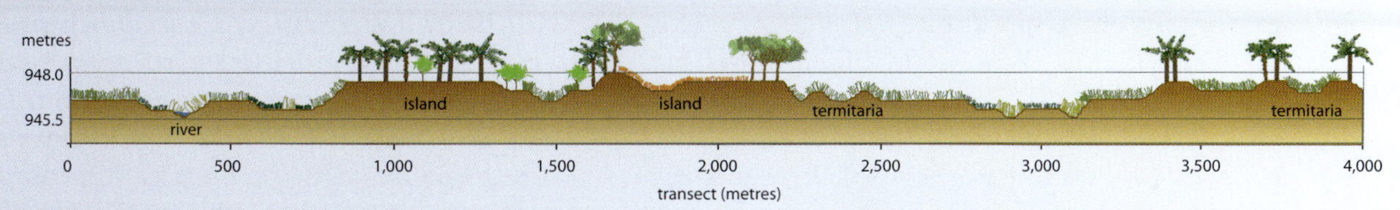

Figure 26 | *This diagram provides an indication of the extent of local variation that exists in elevation, and consequently, in habitat types. This transect is along a distance of 4,000 metres over which elevations changed by no more than 3 metres!*[6]

jut into the Delta **(Figure 25)**. However, many of these dryland species are absent from the adjacent, more arid savannas because they require higher levels of soil and air moisture, or only grow in areas where groundwater is shallow. They are thus very much associated with the broader wetland conditions offered in and close to the Delta. Prominent examples of these are jackal-berry (*Diospyros mespiliformis*), knob-thorn (*Acacia nigrescens*) and lead-wood (*Combretum imberbe*).

Most plants thus grow in the habitats that are never or seldom flooded. By contrast, large areas of the Permanent Swamps are dominated by one or two species, particularly papyrus (*Cyperus papyrus*) and phragmites (*Phragmites australis* and *P. mauritianus*) reeds.

While the truly aquatic species are limited to areas of permanent or frequent flooding, many species grow in a wide variety of conditions of inundation. In fact, as many as 35% of all the Delta's plants grow in more than one habitat along a gradient ranging between swamp and dryland.[7] Some of the plants adopt different growth forms depending on water levels. For instance, the floating heart water lily (*Nymphoides indica*) has floating leaves when the water is high, but grows as a terrestrial plant when the floodplains dry. Its leaves then form compact rosettes. An important point is that the substantial – and frequent – variation in flooding has led to the evolution

Islands support the highest density of species, especially along their margins where a concentrated mix of species hugs the interface between dryland and wetland. One study found between 20 and 83 species in different sample plots, each of which measured only 7 by 10 metres.[8]

The number of species found in any broad area reflects the diversity of habitats. Far from being a uniform wetland, the Delta is very much a patchwork of habitats, as many photographs in this book show. And so it is the mix of habitats that gives the Delta such biotic diversity. In addition, the variety of habitats contributes to the abundance of life because adaptable species can move from one habitat to another, growing and breeding as each habitat becomes productive when wetted, and then retreating elsewhere as the habitat dries. Much of the mix of habitats is due to tiny variations in elevation **(Figure 26)**. Thus, a depression may be inundated for much of the year, while another area just 50 metres away may have standing water for only two months because it is 50 centimetres higher. And between these two, there might be a more elevated area that remains consistently dry.

Overall, the diversity and patchiness of habitats progressively increases from the Panhandle towards the Occasional Floodplains, reflecting both the increasing of islands in the more ephemeral areas (*see page 57*). These patterns of variety and the dynamic nature of habitats have several implications for the Delta's conservation. One is that looking after the Delta's biological wealth cannot be concentrated on small areas or particular habitats. Second, the variety of habitats stems from switching and seasonal pulses of water, and so anything that reduces the variation of flooding would lead to a loss of diversity. And third, plant communities may often be in a state of flux, called succession by botanists, as the supply of water waxes and wanes according to levels of inflow and distribution of water. Conservation measures should thus not aim to preserve a particular condition, other than to allow conditions to change naturally! Mechanisms that cause variability need to be preserved.

Reaching out across the water towards sedges and water lilies is this attractive legume *Aeschynomene fluitans*, probably one of many plants in the Delta that fixes nitrogen.

Processes of change that divide the Delta into its assorted habitats operate over a range of time scales, from decades upwards to tens of thousands of years (*see page 25*). Plant species in the Delta have thus had to track and adapt to frequent changes in habitat conditions. Consequently, we might expect many of the Delta's plants to be so specially evolved to local conditions that they are now endemic, and thus occur nowhere else. Adding weight to this expectation is the Delta's perceived isolation as a wetland, which should lead to its plants being genetically separated from those in other wetlands.

However, all the Delta's plants are found elsewhere in southern and central Africa, especially in wetlands to the north in Zambia.[9] Similar distributional patterns are found for animals, most of which also occur in wetlands associated with the Zambezi River.[10] These broader distributions reflect two features of the Delta and its biota. The first is that the Delta is not as isolated as we think. In the past it was part of much larger wetlands that were connected to those of the Zambezi (*see page 24*)[11], and the Selinda Spillway still provides intermittent connections between the Delta and the Linyanti Swamps. And from the Linyanti, there is continuous swamp all the way up the Kwando River (which is more an extremely long marsh than a river) to the wetlands from which both it and the Zambezi River drain.

Secondly, aquatic species in the tropics and sub-tropics have evolved to be highly mobile because freshwater wetlands are often ephemeral. This is true at time scales of seasons, decades or centuries, depending on the reliability of water sources that fill the wetlands. Species that depend on wetlands thus need to develop ways of moving to patches of water as they become available, and leaving when they dry up. Plants use a number of strategies to spread themselves, for example, by having seeds that float, seeds carried by animals in their guts or as attached burrs, seeds that germinate after varying periods of desiccation, and seeds and bulbs that survive long dry periods. In addition, some plants have rhizomes and stolons that creep along under the ground, while aquatic species have flowers above water to enable pollination between plants over wide areas.

Habitats

The distribution of major habitats in and around the Delta is shown in **Figure 27**. This map presents a neat demarcation of labelled categories, and it is broadly valid in reflecting where different habitats predominate. However, demarcations on the ground are far from simple because there are also many sub-groups characterised by more subtle assemblages of species. Moreover, quite different habitats are often found cheek by jowl because of the local variation in topography and inundation **(Figure 26)**. And habitats change when water flows shift from one area of the Delta to another. For example, the swamplands that predominated in the western Delta turned into dry grasslands when the Thaoge River dried up over a hundred years ago (*see page 50*).

Studies by several plant ecologists have all shown that hydrology is the key factor that determines how vegetation is structured into habitats **(Figure 28)**.[12] Three aspects of hydrology stand out: the frequency of flooding, its duration and the depth of water. These are of course often correlated, so that places that are frequently flooded also have the deepest water for the longest periods. Nutrient availability is also important, but less is known about the role of nutrient conditions because these are harder to measure and evaluate. Other significant impacts are from soil salinity (*see page 78*), fire (*page 84*) and foraging by herbivores (*page 87*).

Figure 27 | *The major vegetation types of the Delta. In addition, there are the Island and Rainwater Pan habitats which are too small and scattered to map at this scale.*

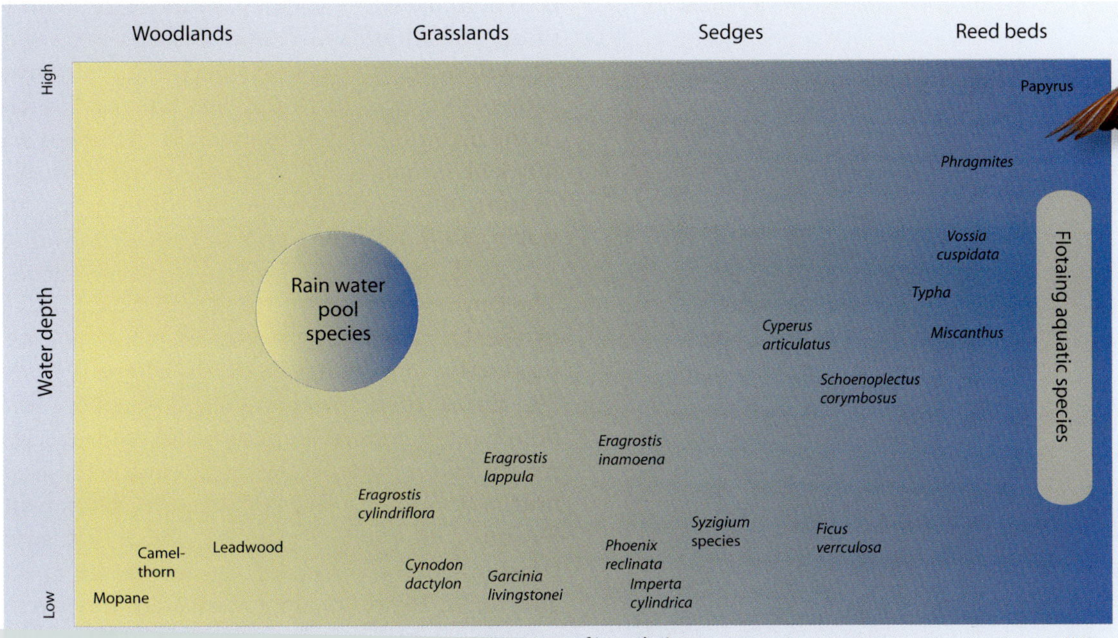

Figure 28 | *The broad relationship between flooding and habitat types, and the approximate preferences of various common or conspicuous species within the realm of flooding frequency (x-axis) and depth (y-axis).*

1. Permanent Swamps

While water levels may drop temporarily to ground level in the Permanent Swamps, they do not drop below its surface. This is a key factor for plants in this habitat, since all are true aquatic species. Four plants are especially abundant and domineering along the margins of the channels and in their immediate back-swamps: papyrus, the two phragmites reeds (*Phragmites australis* and *P. mauritianus*) and the swamp grass (*Miscanthus junceus*). Their growth and abundance is largely determined by water depth, flow rates and their ability to capture nutrients from the passing water. Thus, tall papyrus and phragmites reeds dominate the deeper waters of the Panhandle and upper alluvial fan, while swamp grass is abundant in the shallower lower reaches in the alluvial fan. The plants cannot encroach into channels as long as water flows remain strong, and growth is also limited by their ability to take up soluble nutrients from the water.

Few plants grow in the lagoons or lakes because of their depth, but their perimeters have communities similar to those of the shallow back-swamps. Once dislodged from the bottoms of the lakes by decomposition and methane production, blocks of peat may rise to the surface as floating sudd mats which are then colonised by plants such as *Pycreus nitidus*, *Websteria confervoides*, *Cyperus pectinatus* and *Oxycaryum cubensis*.

The four species lose their dominance away from the main channels and back-swamps where the water is shallower and nutrients from inflows of the Okavango River are less available. These more open waters are occupied by a greater variety of plants, most of which grow as emergent, floating and submerged species. Examples are the bulrush (*Typha capensis*); the sedges *Pycreus nitidus*, *Cyperus pectinatus* and *C. articulatus*; blue water lily or tswii (*Nymphaea nouchali*); the floating heart lily (*Nymphoides indica*); *Brasenia schreberi*; *Ottelia* species; and *Potamogeton* species.

Although some plants are rooted in sandy substrates, others grow on layers of peat that are widespread in the Permanent Swamps. The peat consists largely of decomposing organic plant matter but may also contain considerable proportions of clay. This is particularly true in the Panhandle where much of the suspended clay load is filtered out of the inflowing Okavango River water. Indeed, inorganic sediments may comprise 60-85% of the peat in the Panhandle, compared to less than 15% in peat further downstream in the alluvial fan. The densest layers of peat form beneath phragmites reeds.

No other species characterises the Permanent Swamps as much as papyrus, especially where it is so dominating that it may comprise up to 90% of all plant biomass. This is perhaps the main ecosystem engineer of the Delta because papyrus forms large areas of peat, helps filter out much of the incoming clay, uses much of the inflowing nutrient resource, and is largely responsible for causing major changes in the distribution of water and habitats (*see page 51*).

The Permanent Swamps are the most stable areas of the Delta. Although rates of plant biomass production are higher than elsewhere **(Figure 29)**, almost all the production is through the few dominant perennial papyrus and reed species which take up and retain most nutrients. Few nutrients are thus available in the Permanent Swamps to other organisms which are also constrained by a lack of oxygen in the peat and water. However, the swamps are important refuges for fish before the annual floods. They then leave the swamps for the temporary

The swamp grass (*Miscanthus junceus*), here surrounding the chick and nest of a saddle-billed stork, usually occurs where seasonal changes of water levels are less than 30 centimetres. It thus replaces papyrus and phragmites in the shallower areas of the Permanent Swamps.

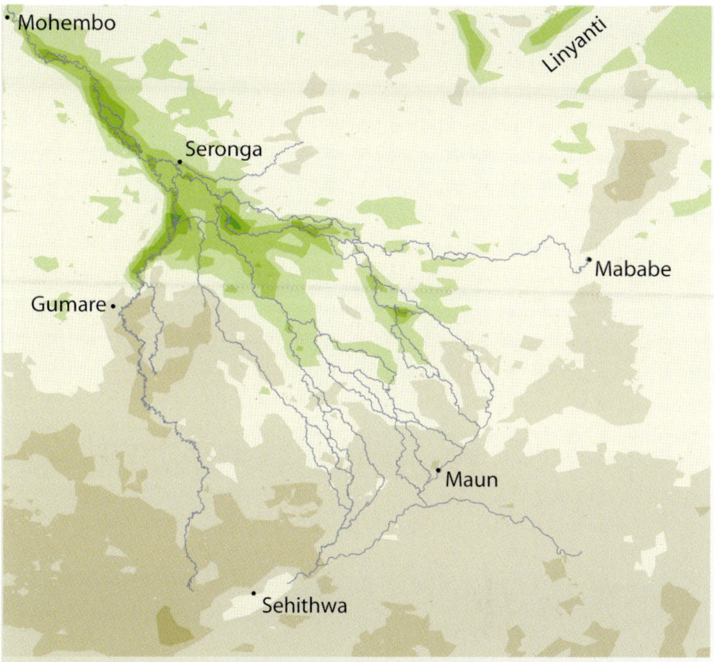

Figure 29 | *A measure of average plant production over 16 years, which shows that most biomass is produced in the extensive papyrus and reed beds of the Permanent Swamps.*[13]

In addition to their upright stalks, phragmites reeds have stems called stolons that grow below the ground where they [illegible] sprout after fire, enabling this vigorous species to recolonise burnt areas rapidly.

Utricularia gibba, one of 12 carnivorous plants in the Delta. These carnivores overcome shortages of nutrients by obtaining them from animals which are trapped and then digested in the small bladders shown immediately above. Species of the genus *Utricularia* have attractive flowers and are most abundant. But the Delta is also home to *Aldrovanda vesiculosa*, which is rare throughout the world.

Papyrus (*above*) is not rooted in the sandy substrate of the Delta, but instead grows and reproduces on layers of peat. This is why it floats so easily on gradually rising water levels. The wide spacing between its stems allows water to percolate through to the back-swamps.The slowing of water flow in blocked channels enhances the growth of papyrus, which then soon invades and further clogs the waterways. But strong flows of water can also break off chunks of papyrus which float off as sudd. If the sudd becomes lodged elsewhere, new communities of papyrus form downstream. Such propagation is very important, since papyrus only reproduces vegetatively. Papyrus also sprouts new growth after fire, as long as their rhizome roots have remained permanently submerged.

The Permanent Swamps typically consist of rather uniform expanses of reeds and papyrus broken only by meandering channels and lakes or lagoons of various sizes. The channels and lakes are often connected by tunnels maintained by hippos (*see page 42*).

floodplain grasslands during the floods to capitalise on the terrestrial nutrients that are mobilised there. And when the flood waters recede, the fish swim back upstream to the Permanent Swamps. Sandbanks along channels in the permanent Panhandle are the most important nesting grounds for crocodiles and African skimmers (*see page* 96).

2. Seasonal Swamps

Whereas the majority of plants stand tall in the deep waters of the Permanent Swamps, most of those in the shallower Seasonal Swamps are shorter. And unlike the evergreens in the Permanent Swamps, many die back when the water dries, to sprout new stems and leaves only when they are next inundated. The key characteristic of this habitat is that flooding is sufficiently regular and deep to support the growth of all life forms of aquatics, especially submerged species. But it is really the sedges that characterise the Seasonal Swamps, setting them apart from the reeds of the Permanent Swamps, the grasses of the Occasional Floodplains, and the woody trees and shrubs of the Drylands **(Figure 28)**.

The arrival of seasonal floodwaters sets in motion a burst of production, which starts when the first microbes are rehydrated. They begin decomposing organic debris, while germinating plants and the regrowth of those that died back in the dry season soon follow. A surprising contest for nutrients that have lain dormant in the terrestrial sediments also begins. As the first bacteria and fungi are revitalized by the water, they start to decompose wetted detritus, rapidly depleting the oxygen in the water to do so. Anaerobic conditions develop, forcing the microbes to use different molecules to metabolise energy, in a chemical process known as reduction.[14] Life is hard without oxygen, and the tough community of different microbes function slowly because using molecules other than oxygen for respiration is a protracted, inefficient process.

The molecules reduced by the microbes for their respiration are also those that plants require as nutrients. Moreover, some molecules become volatile as a result of reduction and then evaporate into the atmosphere. To counter the shortage of nutrients, plants pump oxygen down to their roots, creating micro-atmospheres of oxygenated rhizospheres around their root hairs. The oxygen is for the use of other microbes which congregate around the rhizospheres and use it to oxidise molecular nutrients into forms that the plants can indeed use.

The Seasonal Swamps support a diverse assemblage of short-lived sedges, grasses and aquatics. The mix of [illegible] one place depends strongly on the frequency, depth and duration of flooding, so that grasses [illegible] grass and aquatics are more abundant

While the flooding lasts, the competition for nutrients is fierce and rates of production are rapid, since all plants and animals need to maximise their production within the limited and unpredictable period of inundation. The gist of the competition is for nutrients which can only be obtained from inert detritus when the seasonal floodwaters arrive. While decomposition of the detritus leads to a loss of oxygen, the mass of resulting bacteria and fungi feeds swarms of zooplankton, fish and other animals (*see page 100*). When the waters recede, much of the plant biomass produced during the flooding returns as desiccated detritus to the soil, where it will lie until the Angolan water next arrives in the Seasonal Swamps.

The Seasonal Swamps provide such rich grazing that the high abundance of herbivores in the Delta is probably attributed more to this habitat than to any other. What is especially important is the availability of this grazing during the long, dry winter months when forage in the surrounding Drylands is depleted. Succulent shoots produced by growing grass and from the rhizomes of sedges attract buffalo and zebra within weeks of the floodwaters arriving. They forage in shallow water, while impala graze the fringes of green grass along the shores, and red lechwe feed in the deepest areas. As the waters recede at the end of the flood season, tubers and rhizomes provide food for many baboons and warthogs, and termites proliferate, actively harvesting dead reeds and grass. Attracted by the nutritious termites, dozens of predators arrive to feast their fill.

3. Occasional floodplains

It is often hard to distinguish the Occasional Floodplains from the Seasonal Swamps, and many ecologists prefer not to draw a firm line between the two habitats. Indeed, the boundaries between them **(Figure 27)** are somewhat arbitrary, and also shift in response to longer term changes in flooding frequency. And although the Occasional Floodplains are dry for much of the time, their soils were nevertheless largely formed by alluvial sedimentation.

Grasses with scattered bushes and copses of trees dominate the vegetation of the Occasional Floodplains which are flooded so infrequently that many people would not regard them as floodplains. The most characteristic and common grasses are *Aristida stipoides*, *Chloris virgata*, *Sporobolus ioclados*, *Wilkommia sarmentosa*, *Cynodon dactylon*, *Bothriochloa bladhii*, *B. insculpta* and *Schizachyrium jeffreysii*.

About 220 species of grasses and 120 sedges occur around the Delta. Many of them have small, but intricate flowers, clockwise from top left: *Schmidtia pappophoroides*, *Enneapogon cenchroides*, *Cyperus zollingeri*, *Melinis kallimorpha*, *Kyllingiella microcephala*, *Pennisetum macrourum* and *Dactyloctenium giganteum*.

However, the Occasional Floodplains have a high proportion of plants that occur in no other habitats **(Figure 25)**. Here, regular, annual production depends on rains that fall locally. These features set the Occasional Floodplains apart from the Seasonal Swamps which receive regular pulses of floodwater and share many species with the Permanent Swamps. The Seasonal Swamps also have many more sedges than the grassy Occasional Floodplains.

Since rainfall is usually limited, rates of growth and production in the Occasional Floodplains are generally over many years are mobilised by the water, and production then booms. Large flocks of birds, swarms of insects, herds of mammals and schools of fish are attracted to these productive events. Not only do the animals thrive on this bonanza, but these uncommon episodes of surplus and bounty enable many organisms to reproduce rapidly and prolifically. It is very likely that much of the Delta's biological wealth is indeed produced on the Occasional Floodplains.

4. Islands

The role of island plants in helping to maintain the Delta as a freshwater system by pumping water into the atmosphere was described in Chapter 4 (*see page* 55). The islands also contribute a great deal to the Delta's overall biological diversity, since the estimated 150,000 islands support the most diverse assemblage of species. This is a product of the close combination of terrestrial and fringing aquatic communities and the considerable changes in soil chemistry between the margins and centres of islands. As a result, quite different plants occupy different soil zones.

Changes in soil chemistry across islands are the result of water transpiration by plants, particularly the riparian trees. By drawing up and transpiring water from beneath the ground, a gradient is formed of increasingly concentrated groundwater from the edges to the centres of islands. Calcium and magnesium precipitate out as carbonates on the fringes where the deposits build the slight ridges to be seen around the edges of islands. These precipitates and the gradual accumulation of organic debris cause the islands to grow in size. Sodium, by contrast, precipitates in the middle of the islands where it often forms a salt crust. This crust of sodium carbonate is poisonous to most plants, so the island centres are mostly barren except for a few salt-tolerant species, such as the grass *Sporobolus spicatus* and the ivory palm (*Hyphaene petersiana*).

The fringes of freshwater-loving riparian woodland vary in width from one island to the next. Some are only a few trees and shrubs wide while others extend up to 200 metres in breadth. The commonest and biggest trees in

The biological diversity of island vegetation is valuable in a broad sense, but island plants also provide a suite of resources that support many of the animals in the Delta. For example, grazing and browse is available for herbivores and fruit for frugivores, while many birds use the islands for nesting and roosting. And of course, islands form oases for the myriad of insects, spiders, reptiles and other animals that simply prefer to stay dry.

The variety of species found on an island is broadly related to its size. Thus, the smallest islands are covered by a handful of trees and shrubs, while on the larger, older islands concentric rings of different riparian communities surround zones riparian woodlands are knob-thorns, lead-woods, jackal-berries, sycamore figs (*Ficus sycomorus*), water-berries (*Syzygium cordatum*), wild date palms (*Phoenix reclinata*), large fever-berries (*Croton megalobotrys*), mangosteens (*Garcinia livingstonei*), rain trees (*Philenoptera violacea*), sausage trees (*Kigelia africana*) and marulas (*Sclerocarya birrea*). Beneath and around the trees are diverse communities of grasses and many broad-leafed shade herbs.

5. Rainwater Pans

Although they cover a tiny area of the Delta, Rainwater Pans are disproportionately important because they contain rich accumulations of nutrients and are often centres of biological production. And since they are usually filled by summer rain before the later spread of Delta floodwaters, the Pans provide water to herbivores that have moved into the Drylands to graze fresh pastures that sprout after summer showers. The waters and bottoms of the Pans are also much more clayey than other areas of the Delta. Their most characteristic species are *Echinochloa crus-galli*, *Lemna aequinoctialis*, *Portulaca oleracea*, *Tragus berteronianus* and *Chloris gayana*.[15] About one-third of the plants found in the Pans also occur in the Occasional Floodplains **(Figure 25)**, which is indicative of the ephemeral nature of both habitats where so many species lie dormant until they are activated by water.

The number and size of Rainwater Pans depend on rainfall, as does the length of time that they offer water to wildlife. Their special value lies in the fact that the Pans often supply water in places that are far from sources of Delta water.

6. Dryland woodlands

The three types of woodlands surrounding the Delta are associated with different soils **(Figure 27)**. Thus, the broad-leafed deciduous Kalahari Woodlands to the north are on aeolian sands, the Acacia Woodlands to the west and south grow on deep alluvial sands, while Mopane Woodlands to the east are rooted in alluvial sediments that are more clayey and shallow.[16] The alluvial soils of the Mopane and Acacia Woodlands were formed during much wetter periods when the floodplains of the Delta were several times their present size and the Makgadikgadi mega-lakes extended over huge areas of Botswana (*see page 25*).

The composition and structure of the woodlands is rather homogenous in many areas where just a handful of tree species dominate the landscape. However, there is also much local variety, particularly where soils have been changed by water flows and sedimentation. For example, the Kalahari Woodlands (characterised by red syringa (*Burkea africana*), Zambezi teak (*Baikiaea plurijuga*), and Angolan teak (*Pterocarpus angolensis*) are interspersed with inter-dune valleys where acacias predominate on soils that are more clayey. Similar mixes are to be found in Mopane Woodland where uniform stands of mopane occupy higher ground while mixed communities of camel-thorns, jackal-berries, lead-woods and rain trees grow along old, low-lying drainage lines. Large patches of shrubland savanna dominated by shrubs (mainly species of *Grewia*, *Rhus*, *Maytenus*, *Ximenia*, *Croton* and sickle-bush [*Dichrostachys cinerea*]) mixed with occasional trees (*Acacia*, *Combretum*, and Shepherd's Bush [*Boscia albitrunca*]) occur where soils are sandier than most of those in the Acacia woodlands.

The Acacia Woodlands are characterised by camel-thorns, which have deep roots to tap the groundwater, while mopane have extensive shallow root systems to take advantage of even the smallest local rainfall event. The clayey soils preferred by mopane also hold more water than the sandier soils in which camel-thorns grow.

Mopane Woodlands are relatively tall and dense close to wetlands, while on drier areas they are more open with most trees of uniform height. There are also large areas of shrubland known locally as *gumane* where the mopane shrubs appear stunted, perhaps by fire, frost, browsing by elephants or some feature of the soil.

The woodlands provide valuable pastures and browse, especially highly nutritious seed pods, for herbivores associated with the Delta. Without the woodlands there would be few impala and kudu, for example, and it is probably the large areas of mopane wood- and shrubland that allow the huge elephant populations to exist (*see page 106*). Pastures in the Acacia and Mopane Woodlands are substantially more productive because the nutrient content of their alluvial sands and clays is higher than that of the wind-blown sands of the broad-leafed Kalahari Woodlands.

7. Lake Ngami and the Mababe Depression

Sediments very rich in nutrients have accumulated in these low-lying extensions or overflows of the Delta as a result of their repeated flooding and drying over hundreds of thousands of years (*see page 25*). Both have been dry for most of the time

Thousands of hectares of Lake Ngami were covered by very dense stands of *Datura ferox* early in 2009 (*left*). An extensive community of umbrella-thorn trees had grown up in a zone around Lake Ngami that was last flooded in the late 1970s. Flowers dropped by the trees produced carpets of yellow when the exceptional flood waters arrived in the second half of 2009 (*right*).

like as much as in the past.[17] As elsewhere, the structure and composition of plant communities in Lake Ngami and the Mababe Depression are strongly influenced by inundation. Those in the lowest, frequently flooded areas consist of grasslands when dry, and a variety of sedges, reeds and floating plants when wet. Algae are particularly abundant because of the rich supply of nutrients in the sediments. However, other than the hardy couch grass (*Cynodon dactylon*), little grass cover remains in the bed of Lake Ngami because of very high grazing pressures by cattle. The removal of native species has enabled substantial stands of invasive pioneer plants, such as *Datura ferox* and *D. stramonium*, the burweed (*Xanthium strumanium*) and *Sesbania* species to take hold. Cattle are absent from Mababe so that the grasslands, dominated by nutritious buffalo grass (*Cenchrus ciliaris*) and couch grass remain in good condition here.

The higher zones around both depressions are wooded, with umbrella-thorn (*Acacia tortilis*), camel-thorn (*Acacia erioloba*), buffalo thorn (*Zizyphus mucronata*) and lead-wood being conspicuous and abundant.

DIATOMS

Early studies focused on documenting what species occur in the Delta, and over 180 species of diatoms have been identified so far.[18] More recent research has sought to understand the preferred ecological conditions of different species and to evaluate the potential for using diatoms as indicators of water quality and the overall health of the Delta. These latest studies found that the Delta's diatoms are characteristic of species that prefer water with low alkalinity, that most species also occur elsewhere in the world and Africa, and that different communities of diatoms inhabit different areas of the Delta. Factors that appeared to influence the distribution and abundance of diatoms included water chemistry – such as levels of calcium, silica and alkalinity – and the duration of flooding. A few rare species of diatoms were found, but more work needs to be done to determine if they are endemic to the Delta or not.[19]

None of the species shown on the facing page have English or other common names, and they are very small, ranging from less than 0.01 to 0.3 of a millimetre. The plants in these images have thus been enlarged about 1,000 times.

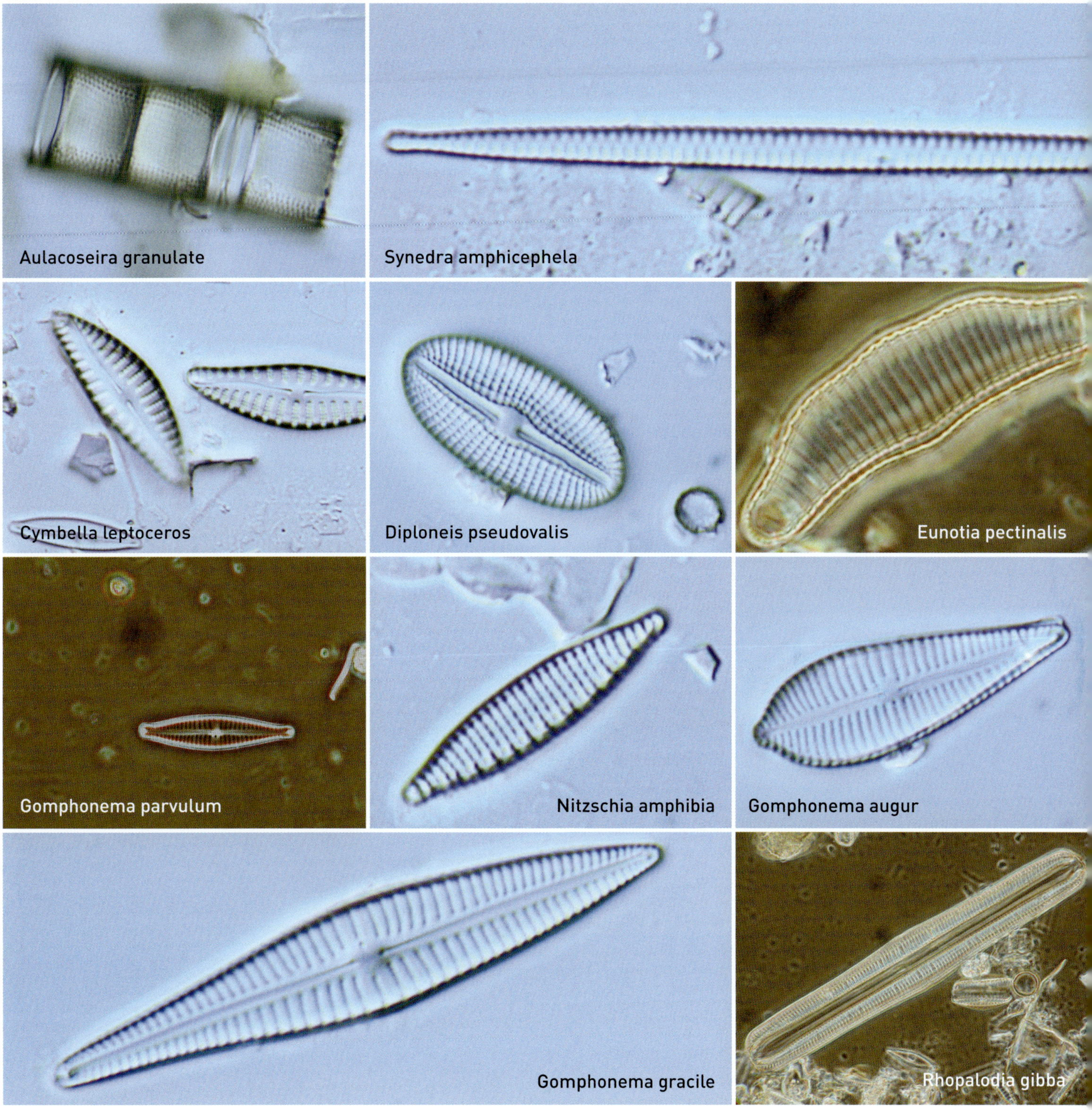

Aulacoseira granulate
Synedra amphicephela
Cymbella leptoceros
Diploneis pseudovalis
Eunotia pectinalis
Gomphonema parvulum
Nitzschia amphibia
Gomphonema augur
Gomphonema gracile
Rhopalodia gibba

Diatoms are tiny algae found in most aquatic environments. Each plant consists of a single cell, the walls of which are composed of silica sculpted in intricate structures that are unique to each species. As photosynthetic algae, diatoms are primary producers and add oxygen to the Delta's waters. They also play important roles in cycling chemicals, particularly silica, and in contributing to the diversity and ecological resilience of the Delta. An average litre of water in the Delta may contain from tens to thousands of individual di

Fire, herbivores and people

The Buffalo Fence separates areas of the Delta used for wildlife management from those where cattle and other livestock are held (*see page 78*). It is a small, simple wire fence but its position is easily visible from space because of differences in vegetation on either side of the barrier (see, for example, the book cover and the image on *page 44*). More specifically, the differences are in the density of trees and shrubs, so the darker areas on the images have more woody plants whereas grasses are more abundant in the paler areas. Most people agree that these differences are a consequence of different grazing pressures by livestock and/or wildlife. But just how this results in more or less grass, or greater or lesser woody plant cover, is debateable. The debate revolves around several possible consequences of grazing.

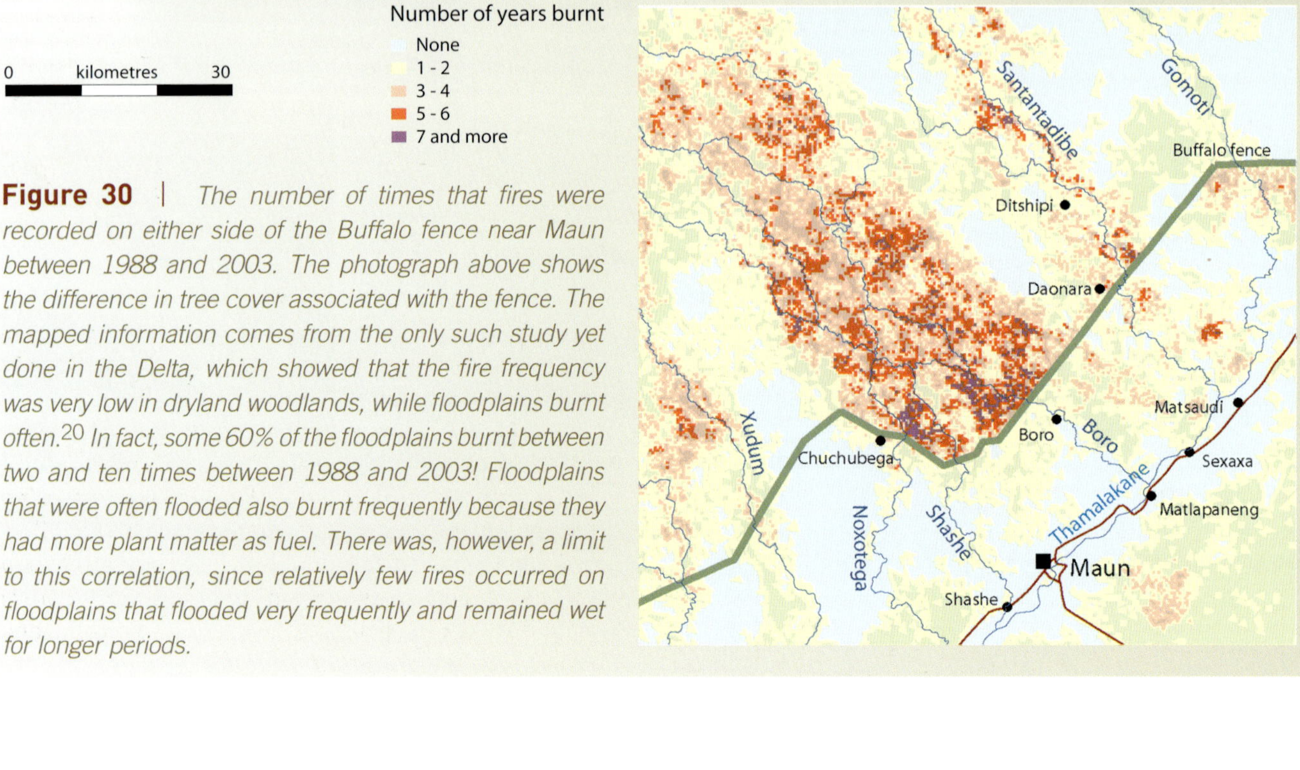

Figure 30 | *The number of times that fires were recorded on either side of the Buffalo fence near Maun between 1988 and 2003. The photograph above shows the difference in tree cover associated with the fence. The mapped information comes from the only such study yet done in the Delta, which showed that the fire frequency was very low in dryland woodlands, while floodplains burnt often.[20] In fact, some 60% of the floodplains burnt between two and ten times between 1988 and 2003! Floodplains that were often flooded also burnt frequently because they had more plant matter as fuel. There was, however, a limit to this correlation, since relatively few fires occurred on floodplains that flooded very frequently and remained wet for longer periods.*

The great majority of Delta fires are started by people, mainly during May before the annual floods and in September before the summer rains and first associated lightening.[21] Some are accidental while other fires set on purpose often run away out of control. A host of reasons have been offered for the setting of fires: to attract wildlife to fresh pastures for hunting and to offer tourists greater chances of seeing wildlife, to improve the quality of grazing for livestock and wildlife, to gain access to fishing grounds, to improve the quality of thatching grass, reeds and papyrus, and to clear land f

A variety of plants are used by residents of the Delta. For example, the rhizomes of blue water lilies (locally called *tswii*) are eaten, while the leaves of *Neptunia oleracea*, *Hibiscus cannabinus* and *Alternanthera sessilis* are cooked as spinach. The last species is also used to make soap and can be used to treat digestive problems. Marula berries are a popular food. The fruit of water-berries are edible, can be used to treat diabetes and make alcoholic beverages. Sap from the ivory or real fan palm (*mokolwane*) is fermented into wine. The bark of large fever-berries is used to poison fish. *Indigofera tinctoria* and bird plum (*Berchemia discolor*) are used to dye baskets. Phragmites is widely used for thatching roofs and making palisade walls. *Mokoro* canoes are hollowed out of the trunks of rain trees and jackal-berries.

Most people appreciate elephants very much, but their rapidly growing numbers (*see page 106*) are cause for concern because woodlands in and around the Delta are increasingly being decimated. Mopane Woodlands suffer the greatest damage since elephants relish their leaves and bark. Trees that have been killed or pushed over provide fuel that intensifies fires.

First, high stocking rates and over-grazing may *limit* the abundance of woody species when herbivores physically damage young woody plants by trampling or browsing them. In addition, grazing may stimulate perennial grasses to grow so vigorously that this, too, constrains the growth of young shrubs and trees. Densely stocked areas then have less shrub and tree cover. Second, the removal of grass by grazers can *encourage* the growth of woody shrubs and trees because more water and nutrients are available to them. High stocking rates then leads to the opposite effect: greater bush density.

A third consequence of high grazing pressures is that they reduce the availability of dry grass fuel for fires which curtails the intensity and spread of burning. Woody shrubs and young trees are therefore not killed or burnt back, while those in ungrazed areas are burnt down each time a hot fire passes over them.

This third effect is perhaps apparent on either side of the Buffalo Fence close to Maun **(Figure 30)**. To the south and east of the fence where cattle are abundant, fires are infrequent (and probably less intense) and there is a dense, bush-encroached cover of young mopane. The same trees are virtually absent a few metres across the fence where hot fires are frequent and grazing pressures are low. Another explanation, which does not invoke burning, is that the woodiness south of the fence is due to a combination of livestock grazing intensity (the second idea described above) and a lack of browsing by elephants.

The association between an absence of fire and bush encroachment is troublesome because grass production decreases as bush densities increase. Pastures for both cattle and wildlife are thus lost. The problem is also hard to reverse, since the chances of having fires hot enough

The waters of the Delta are adorned with attractive flowers, clockwise from top left: *Nymphoides indica*, *Cycnium tubulosum*, *Nymphaea lotus*, *Nymphaea nouchali*, *Utricularia foliosa*, *Ottelia ulvifolia* and *Lagarosiphon ilicifolius*.

increase and fewer grasses establish themselves. Fire is also a problem in the Kalahari Woodlands (*see page 80*) where frequent, intense fires gradually convert the tall savannas into shrubland. Mature trees are killed by the fires as a result of the progressive burning of their stems and bark.The trees are seldom replaced because there are too few fire-free years for young trees to grow tall enough to escape the effects of hot fires. In essence, there are just too many hot fires which rage through the Kalahari woodlands almost every year.

Fires are a problem too, when resources such as thatching grass, reeds, timber and even homes are lost.[22] Burnt landscapes also have no aesthetic appeal for tourists. Furthermore, urgent concerns are often raised about the impact of frequent, hot fires in changing the composition of plant communities, and causing losses of wildlife, water birds and fish production. Good evidence for the nature and severity of these impacts is lacking, however.[23]

One study compared plant and animal production between two adjacent Seasonal Swamp areas, one of which had just had a very intense fire.[24] All the organic debris and peat was removed by the fire. No grasses and sedges grew in the first season when floodwaters returned to the burned floodplain, but its production of algae was substantial. By contrast, production on the unburned floodplain was dominated by bacteria that decomposed plant litter. Populations of zooplankton (*see page 100*) were similar on the two floodplains, but the number of fish fry on the burned area was 15 times lower than on the unburned floodplain.

The reason for this dramatic difference was not clear, and the results need to be confirmed. But it is clear that by removing the grass-sedge-litter matrix, hot fires change the habitat structure and decomposition of organic debris. Their impacts on the functioning of floodplains may therefore be considerable, and this needs more investigation. Research is also needed to study what is widely agreed to be two major benefits of fire in the Delta. The first is the role of fire in releasing nutrients that have accumulated in peat and other organic debris (*see page 59*). Without fires, much of the biological value of nutrients that have built up in the Delta would probably not be available. Secondly, fires appear to play a major role in causing changes to the distribution of water (*see page 51*), thereby adding to the Delta's diversity and allowing some areas to be wetted and others to dry (and therefore, paradoxically, to be burnt).

While fires are perhaps more frequent than before (because most are set by people), it is also obvious that burning is a natural process. Communities of plants, animals and people have endured the costs and enjoyed the benefits of fires in the Delta for time immemorial. There are also costs if fires are too infrequent, for example when the hot, killing infernos consume large masses of plant material that have accumulated over years. The challenge now is to learn more about the effects of frequent, hot fires and to understand how the devastation of burns (as particular events) measures in relation to burning (as a process).

KEY POINTS

1. Plants are important 'engineers' in causing channels to switch direction, sieving out mud and clay, and pumping water into the atmosphere which helps maintain the freshwater of the Delta.
2. About 1,300 species of flowering plants have been identified in the Delta. Most are grasses, sedges, asters and daisies, and bean and pea pulses.
3. Most plant species occur in habitats that are never or seldom flooded, whereas areas that are permanently flooded are dominated by a handful of species, particularly papyrus and phragmites reeds.
4. The number of species found in any broad area reflects the diversity of habitats, of which there are many in the Delta. It is thus the mix of habitats that gives the Delta such biotic diversity.
5. Most of the Delta's plants are found elsewhere in southern and central Africa, especially in wetlands to the north in the upper Zambezi River Basin.
6. The frequent fires have a variety of effects. On the one hand, plant resources, wildlife, fish production, and nutrients may be lost. But fires also release nutrients held in peat and other organic debris, and cause changes to the distribution of water which add to the Delta's diversity.
7. Conservation should not aim to protect particular habitats or conditions, but rather preserve processes that allow for change and plant diversity.

6

Animals: shifting consumers

More tourists are drawn to the Delta by its animals than by any of the wetland's other spectacular features.

Likewise, this wealth of wildlife also largely attracts and focuses the attention of international conservation groups on the environmental value of the Delta. And for local people, wildlife and fish are important as sources of food and revenue that pay salaries and royalties, and that boost the economy of Botswana. So when it comes to the question of what commands attention, what elicits sympathy, or what generates altruistic donations, wildlife wins hands down. Plants may well provide the habitats and food on which animals depend for their very existence, but they are simply not as sexy as hippos, lions, fish eagles, reed frogs or even damselflies.

Other chapters attempt to provide reasonably comprehensive overviews of various topics: the Delta's history (*Chapter 2*), sources of water (*Chapter 3*), functioning (*Chapter 4*), people (*Chapter 7*) and challenges (*Chapter 8*). Here we take a different approach by focusing on a few select groups of animals, each of which tells a different story about life in the Delta. For example, information on tiny crustaceans, which few people ever see (but may swallow unwittingly!) shows how different species respond to flooding and the roles they play in cycling energy from one level of a food web to another. Birds, more mobile than most creatures, provide glimpses of the challenges animals face in tracking sources of food when floodwaters mobilise nutrients into new cycles of biological production. Dragonflies and damselflies illustrate how life cycles are adapted to different flooding patterns, and show how species in the Delta relate to those in neighbouring areas of southern and central Africa.

These and other perspectives provided in this chapter paint cameos of the varied and fascinating strategies adopted by animals that inhabit this wetland. But before continuing, it is useful to recap on some of the fundamental aspects that define the 'playing field' on which the lives of the animals are played out.

The first fundamental is the great diversity of habitats, which broadly range from drylands (of different kinds) to wetlands (again, of different kinds), to very local levels where a patch of bush grows right next to a pool of water. As with plants (*see page 68*), most animal species occur in the drylands, as the figures in the following table indicate:[1]

	Reptiles	Birds	Mammals
Total number of species	64	444	122
Perennial swamp	7	112	3
Seasonal swamp	5	57	21
Dryland	52	275	110

Over long evolutionary periods, this diversity provides opportunities for specialization so that different species are adapted to different environments. On a day-to-day time scale, the habitats allow animals to enjoy a broad canvas of resources. A young elephant can, for example, indulge on phragmites stolons (*see page 73*) in the morning and mopane bark (*see page 87*) in the afternoon.

Secondly, the landscape is ever shifting over large areas of the Delta, particularly in the alluvial fan's seasonal swamps and occasional floodplains (*see page 44*). As waters come and go, so do food supplies, secure nesting sites, or safe resting places. Opportunism is key to exploiting these changes, and animals need ways of rapidly moving to choice areas, and of escaping those that offer diminished resources or looming danger. They must also be able to locate – and to choose – patches that can offer up the best prospects and rewards, and they need mechanisms to determine the most opportune times to make their movements.

Third, the diversity of species and the astounding numerical abundance of animals in the Delta are very much a consequence of its nutrient wealth. But other factors serve to make this Africa's last Eden.[2] One is the variety and close mix of habitats, as noted above. Another important feature is the availability of a perennial supply of water. Without it, only animals especially adapted to conserving water could exist here. Moreover, life in the Delta is particularly attractive because

Rather little is known about the Delta's frogs, even though many are abundant, brightly coloured and vociferousness. All the species are dependent on water – at least for some part of their life cycle – but some spend long periods underground, waiting for flooding or the filling of rain water pools (*see the photograph on page 39*). Left column, from top: western olive toad (*Amietophrynus poweri*), russet-backed sand frog (*Tomopterna marmorata*), snoring puddle frog (*Phrynobatrachus natalensis*), southern foam nest frog (*Chiromantis xerampelina*), bubbling kassina (*Kassina senegalensis*). Right column, top: Angolan reed frog (*Hyperolius parallelus*), which occurs in a variety of patterns and colours. To the left: long reed frog (*Hyperolius

animals have access to water for extended periods, with the summer rains being complemented by the spread of floodwaters during the winter months. In this Eden, wise humans have played a role by preserving the environment and its wildlife. Without their foresight, the Okavango Delta might now be known for its rice paddies or as a former place of glory, with the water drained to far-away mines and cities.

The wildlife of the Delta is therefore not to be taken for granted.[3] Neither should its pools of nutrients, the pulsed flows of water through Namibia from Angola, and the benefits of good management.

Dragonflies and damselflies

In the tradition of bird-watchers, a growing number of people have taken to watching these colourful insects. Hovering and darting, they settle, the dragons with their ostentatiously outstretched wings, and the damsels with theirs demurely folded. Perhaps because of their conspicuous beauty and the efforts of specialists,[4] more is known about this group (called Odonata) of insects than about any other in the Delta. So far, 99 dragonfly species (which, hereafter, includes damselflies) have been found in the Delta. The highest numbers in the world of ten of these species occur in the Delta, endowing it with special value for the conservation of these populations.

The assemblage of species found in any one area is related to patterns of water availability and flooding. For example, dragonflies around temporary pools are typical of the species found at ephemeral rainwater pools in the dry Kalahari. Not surprisingly, their life cycles are rapid, with adults emerging only 30 days after the eggs have been laid. Species of the permanent swamps, by contrast, require water throughout the year.

The highest number of species occurs where permanent and temporary waters are close together. Assemblages of dragonfly species also change when patterns of inundation alter. For example, only species typical of temporary pools were found on the Thamalakane River when it received intermittent flows during the early 2000s, but these were replaced by more 'perennial' species when the river received water year after year after 2005.

The protracted presence of water from rainfall and flooding allows some species to breed twice in one year. For example, the damselfly *Pseudagrion deningi* is very common between December and March, and then a second generation emerges in late July. Since dragonflies are less abundant in the cooler winter months, reduced competition from other species may be one advantage enjoyed by the second generation.

Of course, the main reason for so many dragonflies being here is the presence of a wetland, although not many people will know that this is because their larvae are aquatic predators. Most of the species found in the Delta also occur in other permanent waters nearby, the closest of these being the perennial wetlands of the Zambezi River and Kwando River catchment. This broad swathe of swamps, rivers, lakes and floodplains in eastern Angola and Zambia (which includes Lake Bangweulu in northern Zambia) is a hotspot of biological diversity and endemism for dragonflies and several other groups, including antelope, flowering plants and fish.[5] The biological resemblances between the Delta and these northern wetlands reflect their similarities as perennial wetlands and historical linkages between them when water flowed more freely between the Okavango, Kwando and Zambezi (*see page 25*).

Crocodiles

These are usually retiring animals, slipping away into the water whenever people approach. Since not much hunting goes on in the Delta, one might assume the population of these giant reptiles to be in good health.

However, their reproduction suffers from considerable disturbance caused by people removing eggs or destroying their nests. In addition, nest sites are altered by burning, trampling by cattle and wakes from motor boats washing over nest sites on sandy beaches. Water monitor lizards often dig up nests to eat the eggs, as well. As a result of all these hazards, only 2% of all eggs laid hatch into young crocodiles.[6]

Dragonflies and damselflies are obvious, abundant and colourful animals of the Delta. For most of the time we see them flying about, seemingly without special regard for the wetlands. But the chief reason for there being so many is that they have to breed in water. In fact, much of their life cycle is spent in water as fierce larvae that eat other animals. Left column, from top to bottom: *Sympetrum fonscolombii*, *Pseudagrion rufostigma*, *Pseudagrion deningi*, *Pinheyagrion angolicum* and right top *Trithemis* [illegible] bottom, left to right: *Ischnura senegalensis* and

Such disturbances would be less of a problem if patches of flat sand suitable for nesting were more widely available. However, few good nesting areas are available, and these are only along certain stretches of channels in the Panhandle **(Figure 31)**. Unfortunately, these areas are generally populated by cattle and people (*see page 115*, and **Figure 37**) and only 43% of areas suited to crocodile nesting is relatively free of human disturbance.

The great majority of crocodiles live in the Panhandle and upper reaches of the Permanent Swamps. Counts over several years indicate that there are about 2,600 crocodiles in the Panhandle, of which close to one quarter are adults. Their food varies according to size, and so the youngest, smallest animals feed on insects and small fish, while the biggest crocodiles take proportionately larger prey. Overall, however, fish form much of their diet in the Delta. In addition to the presence of suitable nesting grounds, the greater abundance of adult fish in the permanent swamps and Panhandle (*see page 101*) further explains the concentration of crocodiles in these areas.

Figure 31 | *Areas of nesting habitat suitable for crocodiles, and remaining areas where they can nest with little disturbance. The map also shows places where African skimmers breed, and the major breeding colonies for water birds in the Delta. The Panhandle has been enlarged to show more detail of the nesting areas.*

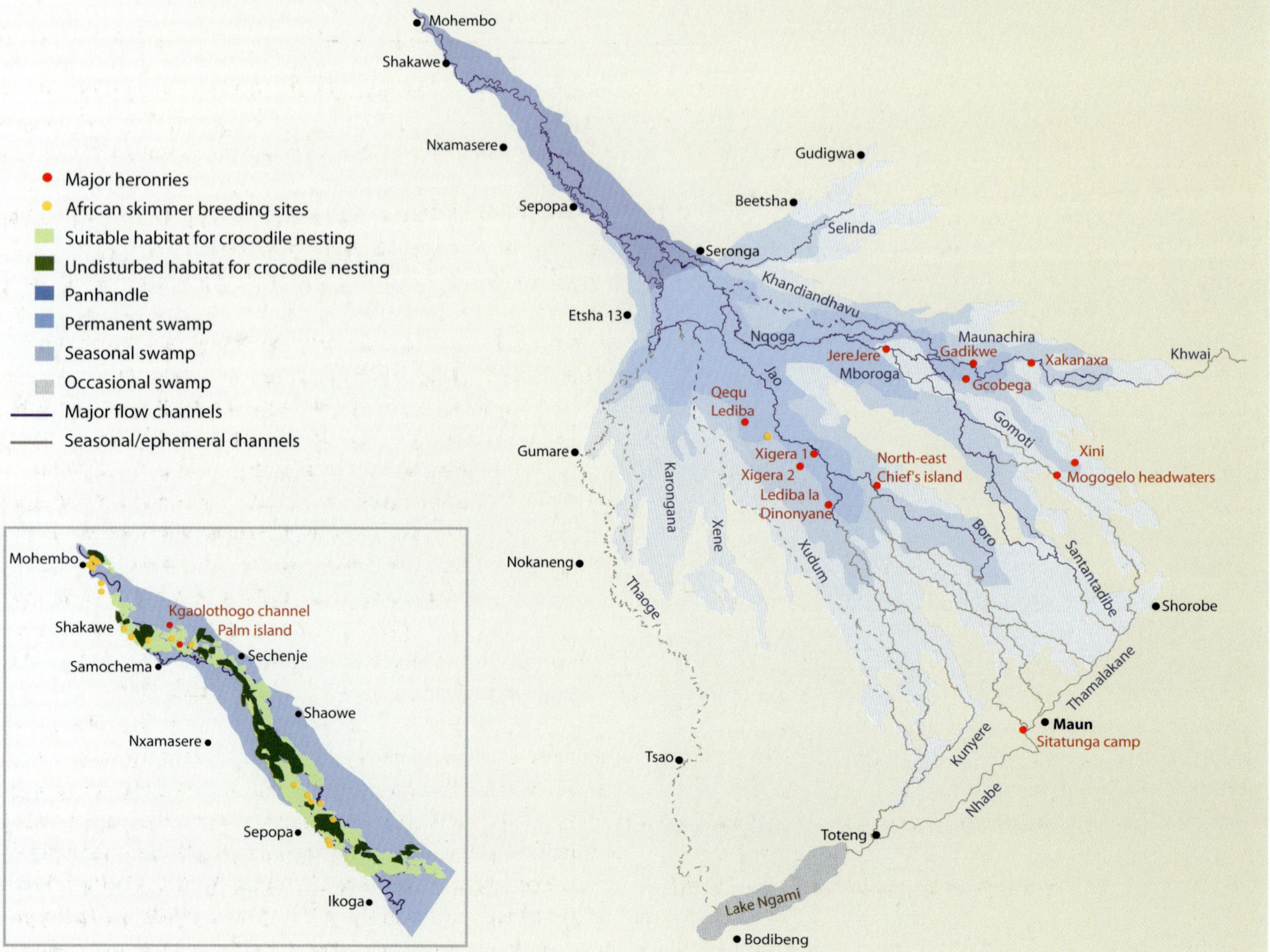

Crocodiles lay clutches of 40 to 80 eggs in nests excavated in flat sandy areas next to large channels. The nests are guarded by females until the hatchlings emerge after about 90 days of incubation. Interestingly, the temperature of the nest determines the sex of young, so that males are produced under warm conditions and females when nests are cooler.

Birds

For much of the time, populations of birds subsist in the Delta, each individual ticking over from day to day. Daily routines start with a period of warming in the rays of the rising sun. Then follows a bout of feeding before a long period of mid-day rest. Another stint of foraging may follow the siesta, before the bird settles down for the night in a secure roost.

But then there are times when the Delta hums with two kinds of productive activity. One is breeding, which boosts the pool of genes, and the other is voracious feeding, which boosts body reserves. Birds of course feed every day, but these are the times when food is super-abundant. Moreover, the feeding events are episodic and they occur in widely separated places. The most important, or at least well known of these spells are:

- emergences of fat-laden termite alates (winged adults), on which vast numbers of birds of many species gorge themselves;
- concentrations of zooplankton and fish fry in seasonal swamps and occasional floodplains which attract flocks of wading egrets, herons and storks; and
- concentrations of large fish trapped in drying pools to which pelicans, storks, fish eagles and other piscivores are drawn (*see the photograph on page 63*).

ability to do so makes the difference between idling and recharging body reserves which enable birds to launch themselves into the other form of (re)production.

While great feeding orgies are often unpredictable, most of the spectacular reproductive events occur annually in well-established breeding colonies. The biggest colonies are shown in **Figure 31**. There are 14 major heronries in the Delta, where spoonbills, pink-backed pelicans, sacred ibises, yellow-billed, openbill and marabou storks, African darters, reed cormorants and several other species can be found breeding. Hundreds of pairs may breed simultaneously, but the total at each colony also varies from year to year. Unfortunately, there are indications that the total number of aquatic birds breeding in the Delta has declined in recent decades. Human disturbance, especially by tourists venturing too close to the breeding birds, is considered to be the main reason for this loss.[7]

These colonies deserve the highest levels of protection, as do the nesting sites of African skimmers and bee-eaters in the Panhandle. The skimmers nest on sandy islands along the main channels, while carmine and white-fronted bee-eaters breed in colonies in vertical sand banks where the main channel has eroded into layers of Kalahari sand. Numbers of breeding skimmers have also declined, largely as a result of the same kinds

An example of how much concentrated food can be available. Up to 5,200 white pelicans have been counted at Lake Ngami, alongside many other fish-eating species.[8] Each of the pelicans would consume just less than a kilogram of fish each day, which means that 5,200 of them would harvest about 5 tonnes of fish daily. White pelicans nesting at the Nata River delta in the Makgadikgadi Pans make return journeys over 600–700 kilometres every day to collect fish trapped in pools in the Delta. They leave the Pans in the morning and return with belly-cans of fish in the afternoon which they disgorge to feed their chicks.

The Delta has very special conservation value for slaty egrets (*left*) and wattled cranes (*right*). The approximately 1,200–1,450 cranes in the Delta constitute the largest single population of this globally threatened species, while an estimated 75-80% of all slaty egrets in the world live in the Delta. This species, too, is classified as globally threatened. In addition, more than 1% of the global populations of 25 other bird species occurs at this Ramsar site (*see page 121*), while another 12 species have more than 0.5% of their populations here. The Okavango Delta Ramsar Site extends across three of Botswana's important bird areas (IBAs) which are the Okavango Delta, Lake Ngami and the Linyanti Swamps/Chobe River.

Of over 450 species of birds recorded in the Delta, about 112 are aquatic birds that feed by diving, wading or swimming, or feed in the vicinity of water. Many of them feed largely on fish which means that they are piscivorous species. This is a selection of fish-eaters. *Top, left to right*: African fish eagle, yellow-billed stork; *middle, left to right*: saddle-billed stork,

Zooplankton

This is the collective name for a variety of tiny animals that float or swim, while phytoplankton is the equivalent name for small floating plants, many of which are algae and diatoms (*see page 83*). The majority of zooplankton in the Delta's waters are crustaceans that belong to groups called Cladocera and Copepoda. Most of them are filter-feeders, sieving out bacteria, small phytoplankton and other minute organisms. Much of what is known about zooplankton in the Delta – and described below – concerns cladocerans, however.[9]

A study of cladoceran diversity found there to be relatively few species in the open waters of channels and lagoons, where their densities were also low. Somewhat more species, some of them in moderate densities, were recorded in the permanent swamps, but the real centre of cladoceran diversity and abundance was in the seasonal swamps.

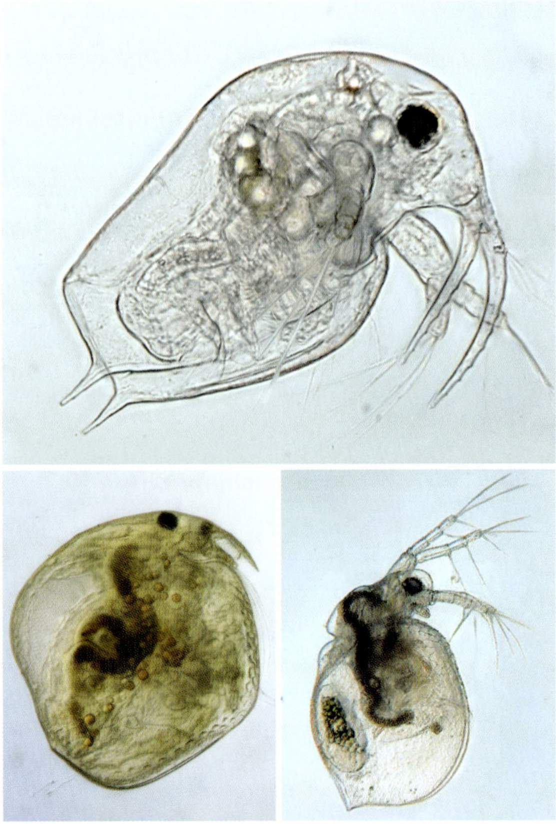

Three of the cladoceran species that are common in the seasonal swamps in the Delta: *Bosmina longirostris* (*top*), *Chydorus sphaericus* (*bottom left*) and *Ceriodaphnia reticulata* (*bottom right*). These names are much longer than the animals themselves, most of which measure less than a millimetre in length or breadth. Cladocerans are commonly called water fleas.

Of course, these animals are not present when the seasonal swamps are dry, but the first eggs hatch soon after the floodwaters arrive. Up to 50,000 cladoceran eggs per square metre have been recorded, all of them lying dormant in the dry sediments. Each egg is enclosed in a thick-shelled ephippium or casing which offers protection during the months – or even years – the egg might lie waiting for water.

Different species in the seasonal swamps hatch at different times, timing their emergence and development to capitalise on the wave of food made available by flooding. Much of the food for cladocerans is provided by vast quantities of phytoplankton, algae and bacteria which decompose particles of plant matter that have lain there since the last flooding.

Moreover, the wealth of production is usually made possible by abundant phosphorous being released into the water, with much of this phosphorous probably coming out of dried mammal droppings. (Accumulations of dried droppings can be as high as 500–700 kilograms per hectare!) In essence, all these constituents of life – the cladoceran eggs, bacterial and algal spores, and then also the plant detritus and phosphorous – remain dormant on the surface or in the sediments until the floodwaters arrive.

Although different species hatch at different times, most of them congregate in the shallowest waters on the edges of the seasonal swamp. This is where nutrient releases are greatest, their food most abundant, and where the cladocerans are best protected from predators, such as fish. Some remarkable densities of cladocerans have been recorded, as many as 100,000 individuals being counted in one litre of water, and in such concentrations that their crustacean odour was distinctly recognisable in the air above the water. Under these conditions, fish, various predatory insects and birds enjoy a veritable feast that does much to replenish their stocks of energy and genes.

Fish

Productive booms in the seasonal swamps (and the occasional floodplains when they get water) do not escape the attention of the Delta's fish, especially those that will benefit from an abundance of zooplankton and lush vegetation.

During dry periods when most water is confined to the Panhandle and permanent swamps, adult fish are likewise limited to standing water in channels, lagoons and reed beds. But as the pulse of inflowing water passes, the fish follow, migrating downstream into areas that are to be temporarily inundated. There, many of the fish spawn as quickly as possible so that the young fry have time to make the most of the seasonal or short-lived food that becomes available in the swamps.[10]

The adults of at least 19 species also remain in these temporary waters to regain body reserves, and to grow larger so that they are more competitive and attractive as mates.

Fish eggs generally hatch within a few days, and the young larval fry then begin the business of feeding and growing as quickly as possible. Most of the fry eat zooplankton and plant matter, since these are the most abundant kinds of food. Indeed, many begin by feeding on cladocerans and copepods when they are small, and later switch to being herbivores when they grow larger.

Under reasonably normal conditions, seasonal swamps hold water for four to six months, during which time sedges and other aquatic plants offer the young fish protection as they grow and grow. For example, species of tilapia can reach lengths of 10–12 centimetres during these few months, each additional centimetre of size reducing their chances of falling prey to predators, such as piscivorous birds (*see page 99*) or African pike.

As a result, growth and survival rates of young fish are directly related to the extent and duration of flooding. Prolonged flooding also provides adults with more opportunities of spawning in the first place, and of growing larger themselves. Thus, in years with little or no flooding, the Delta's fish populations remain static, whereas the highest levels of recruitment occur when flooding is most extensive. The same effects of flooding on fish populations have been found in many other freshwater wetlands.[11]

About 20 fish species in the Delta care for their eggs and fry, such as the red-breasted tilapia (*Tilapia rendalli*) with its young (*top*). This has particular value where predators are abundant and oxygen levels are low, such as in the seasonal floodplains where microbes deplete the water of oxygen (*see page 75*). The aerial foam nests of African Pike (*Hepsetus odoe*) are thus guarded by the parents while the nests provide an oxygen-rich environment for the embryos (*above*). African pike are the predators of the alluvial fan, whereas tiger-fish (*Hydrocynus vittatus*) predominate in the Panhandle. The pike are ambush predators, relying on dense vegetation for cover while waiting for prey, while tiger-fish chase down their prey in open water.

The nests of three-spot tilapia (*Oreochromis andersonii*) have diameters of 70–190 centimetres, and depths of up to 50 centimetres. Males construct the nests after which the females select the best nests and the largest and most colourful males for breeding. The eggs are laid in the nests where they are fertilised by the males and then picked up by the females who hold and protect them in their mouths for up to three weeks.

The famous barbel or catfish run in the Panhandle usually occurs between August and September when many smaller fish return to the permanent beds of papyrus and phragmites after breeding in the seasonal swamps. Packs of sharp-tooth (*Clarias gariepinus*) and silver catfish (*Schilbe intermedius*) move along the main channels, hunting down the smaller fish concentrated along the margins of the channels. These frenzied runs of feeding catfish are followed by tiger-fish, large bream, aquatic birds and other predators which snap up morsels and prey that escape the catfish.

The annual wave of floodwaters which arrives in the lower Panhandle each March and April flushes out water that has been standing in the reed beds for months. This relatively stagnant water contains very little oxygen because it has been used by bacteria that decompose peat (*see page 54*). Great numbers of fish are killed when this water is pushed into lagoons, such as the one at Guma, because there is too little oxygen for them to respire.

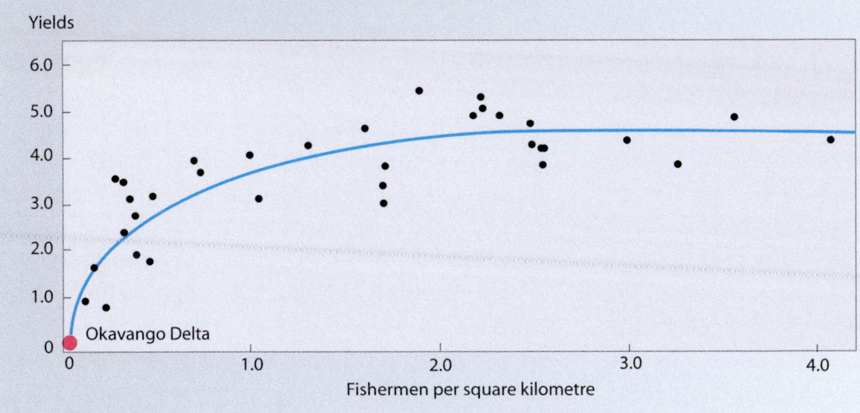

Figure 32 | *Fishing effort (as reflected by the density of fishermen on the x-axis) is generally dependant on yield (y-axis, kilograms per hectare per year), as shown by this analysis for 31 African lakes. Each lake is represented by a dot. Yields and efforts are smallest in the Delta, largely because fish populations are low as a result of the nutrient-deficient waters* (see page 53).[13]

Fishing and fishermen

Most of the 71 fish species in the Delta are too small and uncommon to be exploited, but there are a number of fish that are large and tasty, and that provide a good contest for recreational anglers. The most highly-prized for consumption are tilapia, the three-spot, red-breast and the green-head tilapia (*Oreochromis macrochir*), and they are targeted by commercial fishers who sell their catches to customers in and around the Delta. Anglers are usually after tiger-fish, which have a reputation for offering a sporting fight, as well as the large nembwe (*Serranochromis robustus*).

The number of recreational anglers is not known. Neither is the number of fish nor the weight of fish that they remove, although most people agree that these are not substantial. Commercial fishing operations are concentrated in the northern areas of the Panhandle, where syndicates of fishermen have cold-storage facilities at Samochima, Mohembo West and Ngarange. There are now about 40-50 commercial fishermen in the area, operating mainly from motor boats and using nets.

A much greater number of people fish for subsistence purposes. Most of them are women and young girls who use baskets (*see page 8*), or young boys with hooks and line, or men with spears and barrage traps. The total number of families who fish is estimated to be about 3,000, and most of them live alongside the Panhandle, particularly around the Etsha villages.[12] While fish usually supplement other, more staple foods, such as that obtained from cash purchases and farming, fishing ... valuable safety-net, especially for very poor

fishing during 1996 when all the cattle in Ngamiland were killed as a result of an outbreak of lung sickness (*see page 119*).

The question of whether the Delta is over-fished or under-fished is often debated, sometimes hotly so. Accusations often include claims that there are too few fish to go round or that the Delta is being over-fished. For example, commercial fishermen complain of being excluded from tourism concession areas because there are too few fish in areas in which they normally operate. In turn, tourism operators counter by objecting to fishing by commercial and subsistence fishers in their concessions, because they claim that the local fishermen will deplete stocks available for angling tourists.

Such debate actually revolves more around rights to resources and land (*see page 122*) than the abundance of fish. Nevertheless, fishing off-take has dropped significantly over the past 20 years, from about 3,000 tons per year to less than 1,000 tons.[14] This is perhaps because more people have access to cash incomes and/or because public funds were previously used to subsidise equipment for commercial fishermen and the purchase of their catches.

Over-exploitation certainly occurs locally, for instance in confined lagoons isolated from the easy interchange of fish populations with other waterways in the Delta. However, most research suggests that the Delta's fish are in good health, largely because so few people actually catch fish **(Figure 32)** and because there are large areas where there is no fishing. These are places where fishing is prohibited or that are too far from ... settlements.

Large mammals

Everyone would agree that the Delta is full of large mammals, but why, and how many are there? The first question was discussed in Chapter 4 (*see page 58*) where it was reported that the Delta holds about 10 times more large mammals than would be expected in an area of similar annual rainfall if the underlying soils were also Kalahari sands. And so the key feature that sets the Delta apart in providing sustenance to so many mammals is its nutrient base. Not only do the nutrient rich soils enable forage to grow rapidly and in abundant quantities, but much of the forage is also much more nutritious in quality than it would be on poorer sandy soils.

Answering the second question about how many mammals live here is harder. First, what area are we talking about: the permanent and seasonal swamps (which cover about 6,900 square kilometres) or the occasional floodplains as well (which add another 6,500 square kilometres). Or do we include the larger are of nutrient rich sediments that is now usually dry but which received water flows over tens of thousands of years (*see page 24*)? Another confounding problem is that many mammals move around, spending some months in the heart of the Delta and others far away in the Kalahari sands or around the Makgadikgadi Pans, for example.

Browsing mammals, such as kudu and giraffe (*opposite*), are much more sedentary than grazers, such as buffalo, wildebeest and Burchell's zebra (*above*). Trees tend to produce similar amounts of foliage each year, whereas the

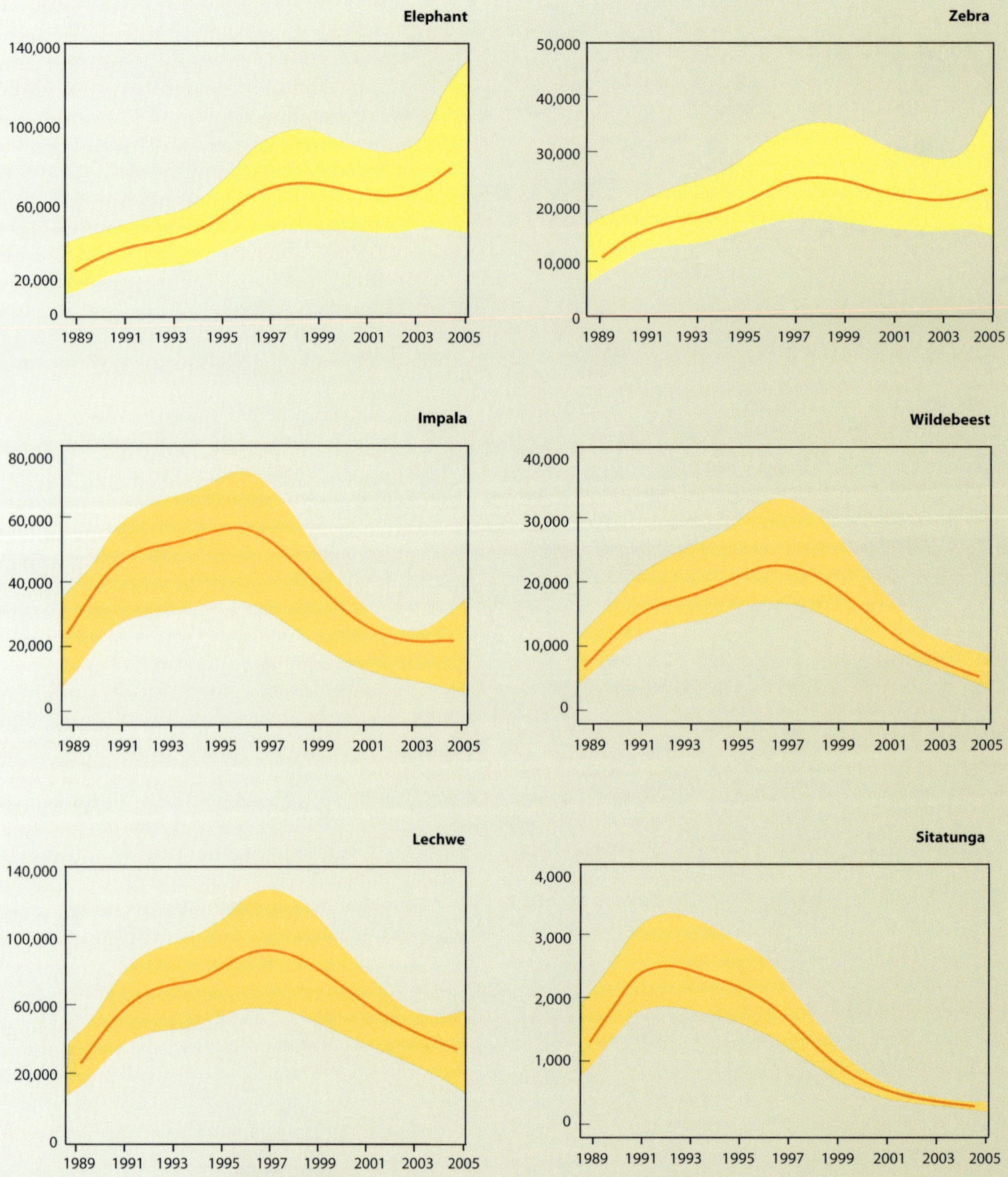

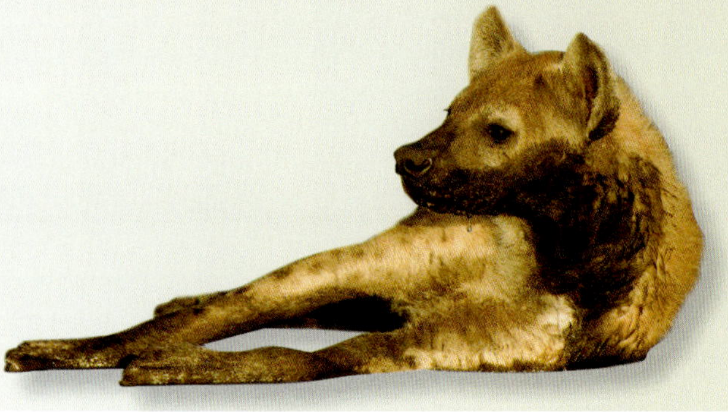

Figure 33 | *Population estimates for large mammals obtained from aerial counts between 1989 and 2005. The thin red lines show the population estimates for each year, while the coloured bands above and below the lines are the 95% confidence limits for the estimates.*

Two spotted predators testing their nerves: a leopard below and large spotted genet above.

Moreover, numbers of large mammals have been changing, and substantially so. For example, data from aerial surveys done by the Department of Wildlife and National Parks over an area of 20,000 square kilometres (which includes all of the proper wetlands of the Delta) indicates that populations of Burchells' zebra and elephant increased several times over a period of 16 years **(Figure 33)**. By contrast, numbers of red lechwe, sitatunga, blue wildebeest and impala rose significantly during the 1990s and then declined by similar orders of magnitude. Rather similar changes were recorded giraffe and waterbuck, while numbers of sable, tsessebe, roan and eland declined more steadily over the same period.

And that begs another question: what causes the population changes? Clear answers are not available, and they may differ from species to species. However, one plausible explanation draws on the consequences of longer term high and low inflows from Angola and the of the Delta have been flooded with deep water for longer periods. This may have limited the extent of floodplain grasslands, thus reducing the availability of nutritious forage for some grazers.[15] This is one hypothesis, and clearly much more investigation is needed to understand factors that influence the population dynamics of these charismatic icons of the Delta.

KEY POINTS

1. The highest diversity of animals is in the drylands that surround the Delta, whereas the permanent and seasonal swamps are the preferred habitats for relatively few species.
2. Populations of fish (and many other animals) are directly dependant on the extent and duration of seasonal flooding because most recruitment occurs in the nutrient-rich seasonal swamps and occasional floodplains.
3. The Delta's populations of crocodiles, African skimmers and carmine and white-fronted bee-eaters breed mainly in the Panhandle where they often suffer from human disturbance.
4. Populations of several large mammals and also of aquatic birds breeding in heronries have declined in recent years, but the numbers of Burchell's zebra and elephant have increased, by contrast.
5. Most of the 99 species of dragonflies recorded in the Delta also occur in wetlands in the headwaters of the Kwando and Zambezi rivers, which is a centre of endemism for various groups of animals and plants.
6. Vast numbers of zooplankton hatch and 'ride waves' of production as floodwaters extend into and over seasonal swamps. In turn, adult fish migrate from permanent waters into the seasonal swamps to spawn where their fry feast on the zooplankton and other abundant food.
7. Per unit area, there are far fewer fishermen in the Delta than in other African wetlands because fishing yields are much lower than elsewhere.
8. Fish populations are relatively low because the nutrient base of the Delta's waters is small. By contrast, large herbivorous mammals are very abundant as a result of sediments and

7 People: predators and protectors

Humankind is both the enemy and guardian of the Delta. Certain people use it, while others abuse the wetland; some people remain on the land, while others move away.

Concerns about increasing pressures being placed on the Delta's natural resources contributed to the reasons for writing this book. Some threats come from far afield, such as climate change and potential developments upstream of the Delta (*see page 130*), but others are due to local processes. And because the Delta matters most to the people who live there, much of the obligation to care for the wetlands falls on local shoulders.

The health of the Delta is thus very much a matter for local people, on whom we now focus. As the chapter will show, this is a sparsely populated area compared to most places in the world. It is also one where people are in rapid transition from traditional lifestyles to others associated with modern economics and commodities. And, with many people being recent immigrants from elsewhere in Botswana and other countries, the population is diverse in character and origin.

The majority of people live on the outskirts of the Delta rather than within the wetlands themselves. For this reason and the fact that most demographic information is aggregated for Ngamiland, many of the statistics presented are for this district.

The people of the Delta have been highly mobile in various ways, and the lives of many continue to change. Making judgments about what is good and what is undesirable is extremely difficult. Charting a course that seems best for today may not hold for tomorrow, and what suits one group of people may be troublesome to others.

Historical perspectives

Associations between people and the Delta started long before it came to the attention of the wider outside world during the second half of the 19th century (*see page 10*). But rather little is known of those earlier times, which are largely shrouded in legend. For example, Tsodilo Hills, which are adorned with over 4,000 rock paintings, are claimed by several peoples, including the Hambukushu, Bugakhwe and Xanikhwe, Ju/'hoansi as their ancestral 'home'.[1]

From archaeological records it is clear that northern Botswana has been inhabited for the past 100,000 years at least, and probably for much longer before that. Evidence of occupation has been found at many sites around the Delta **(Figure 34)**, and we can assume that all areas between these sites were also occupied at various stages.

It is also certain that the Delta's resources, such as fish, game and water, were vital to those who lived nearby, and that people who lived further away would likewise have made excursions to gather food from the wetland.

The Delta and its immediate surroundings have therefore been a home and pantry for much longer than we often imagine. Those who first lived here were hunter-gatherers, and today's so-called Bushmen, San or Basarwa are perhaps their descendants. The populations of the earliest inhabitants would have been small, and they would have moved widely and frequently in pursuit of sustenance provided by wild plants and animals.

Rock paintings at Tsonxhwaa Hill, just south-east of Savute camp. How different was the reasoning of people who painted these images from those who took the photographs in this book? The sets of images were captured in different media (paint, film or digital code) but all conceivably reflect the esteem in which the subjects were held.

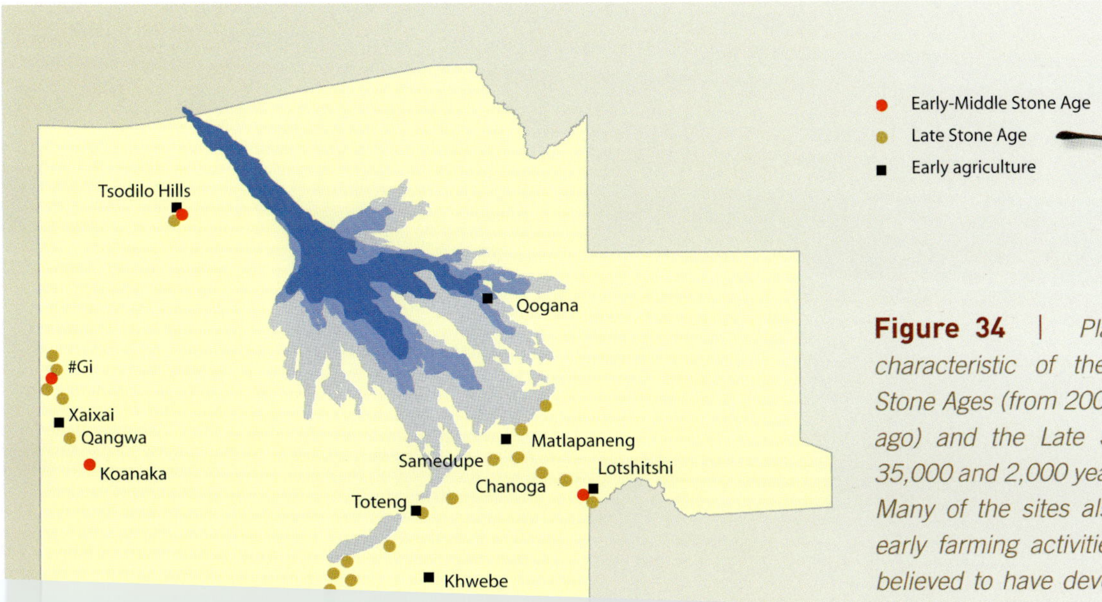

Figure 34 | *Places where artifacts characteristic of the Early and Middle Stone Ages (from 200,000 to 35,000 years ago) and the Late Stone Age (between 35,000 and 2,000 years) have been found. Many of the sites also show evidence of early farming activities and agriculture is believed to have developed here between*

Permanent settlements were established only when livestock and crop farming was introduced by Bantu people. Directly or indirectly, these first farmers were probably the forefathers of the BaKgalagadi, Wayeyi and Hambukushu, Dxeriku, Herero and Tawana, even though some groups appear to have settled around the Delta more recently.[3]

Indeed, the history of most people associated with the Delta is characterised by frequent migration towards newly found natural resources or economic opportunities, or to escape livestock and human disease, conflict, drought or flooding.

For example, the Tawana people initially came on hunting forays to Ngamiland, where they established their first settlements in about 1800. Over the next 100 years, they moved their capital eight times between places such as Kgwebe, Toteng, Nokaneng and Tsao. The last move was in 1915 to Maun.[4] The rinderpest epidemic in 1896 wiped out all the cattle in Ngamiland, while outbreaks of sleeping sickness (transmitted by tsetse flies) led to several settlements in the Delta being abandoned in the 1940s and 1950s.

Conflict frequently played a role. Raids by the Matabele in the second-half of the 19th century forced residents of many settlements to flee. Most Herero people escaped to Ngamiland to avoid extermination during the 1904-1906 German-Herero war in Namibia. And more recently, Hambukushu people were relocated to the Etsha settlements in 1969 and 1970 to evade conflict in Angola.

Exoduses to escape hardships are better documented than movements prompted by the lure of new resources or opportunities, which are more gradual and less dramatic. As we shall see, mobility in pursuit of better livelihoods remains a feature as large numbers of people forsake their rural homes for urban ones.

The first attempt at a systematic appraisal of population size was by Siegfried Passarge in 1898. By counting huts and villages, and estimating how many villages lay beyond his route, he arrived at an estimate of 5,000 inhabitants in Tsau, the then Tawana capital, and between 20,000 and 25,000 in the whole region around Lake Ngami.

And families now often have diverse sources of income. This is another characteristic with a long history that stems from the need to be flexible and resourceful in a land where rainfall and flooding was variable, diseases common, and natural resources spread over large areas, for instance.[5] Family members therefore often had different roles. Their activities during the dry season also differed from those after rain had fallen, or from one year to the next, depending on flooding, access to resources, labour and capital.[6]

The number and spread of people

Official census figures were gathered in the first half of the 20th century by requesting everyone to assemble and be counted at villages of their headmen. This was the basis for estimates of 21,550 people in 1904, and also for figures in 1911, 1921 and 1936, some of which seem rather high or low **(Figure 35)**. Subsequent counts were more reliable, and we can be confident of the population totals that have risen so steeply between 1964 (42,572 people) and the latest estimate of 138,654 people in 2006.

Whilst migrations led to some population increases, most of the district's recent population growth was due to better survival rates, particularly as a result of health

Figure 35 | *The population of Ngamiland has grown rapidly, particularly after the widespread introduction of health services during the 1960s and 1970s. Between 1981 and 1991, the number of people grew by 3.3% each year. This rate dropped to 2.8% between 1991 and 2001, and is now estimated to be 2.1%.[7] This is good news, but only means that the* rate at which more natural resources are consumed each year *is perhaps not as high as it was. Indeed, demands on natural resources continue to rise even if populations remain constant because so many people seek western, consumerist livelihoods, a trend made clear by the rapid growth of Maun's population.*

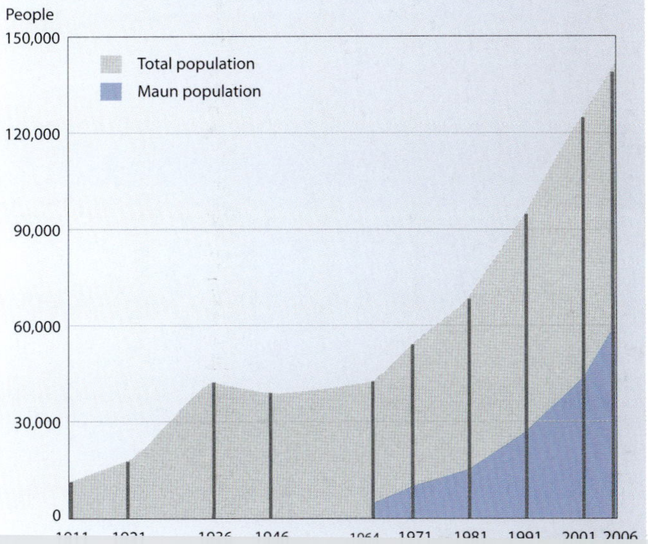

Figure 36 | *Declining birth rates in Ngamiland are reflected by the smaller proportions of young children in 2001 compared to 1991. The age pyramids also show that the swelling population of Maun consists largely of people of working and job-seeking ages.*

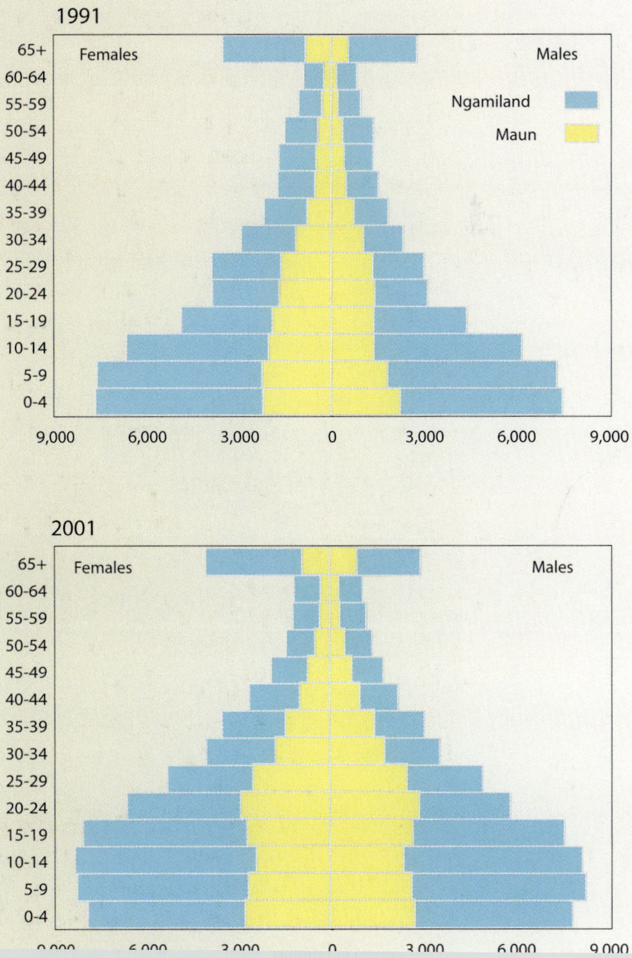

Some idea of the enterprise and bustling activity in the booming town of Maun is provided by these images. Maun was officially a village before gaining formal status as a town in April 2009, when it had about 60,000 residents. Its population has been doubling every 11 years.

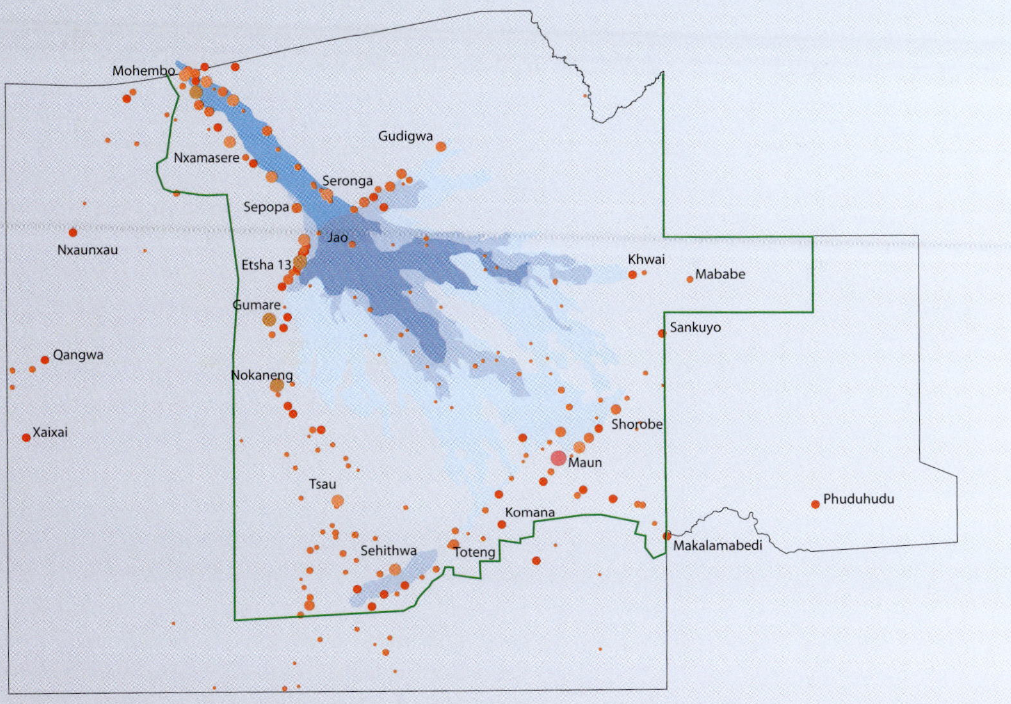

Figure 37 | *The great majority of people live along the western and southern margins of the Delta where most large villages and settlements are located* (top)*. The map of Ngamiland's human footprint below provides an idea of the magnitude of the man-made pressures that are being brought to bear on resources. The measures are based on estimates of population density, the amounts of land transformed for settlements or by clearing for cultivation, and locations of roads which reflect human access to natural resources.*[8]

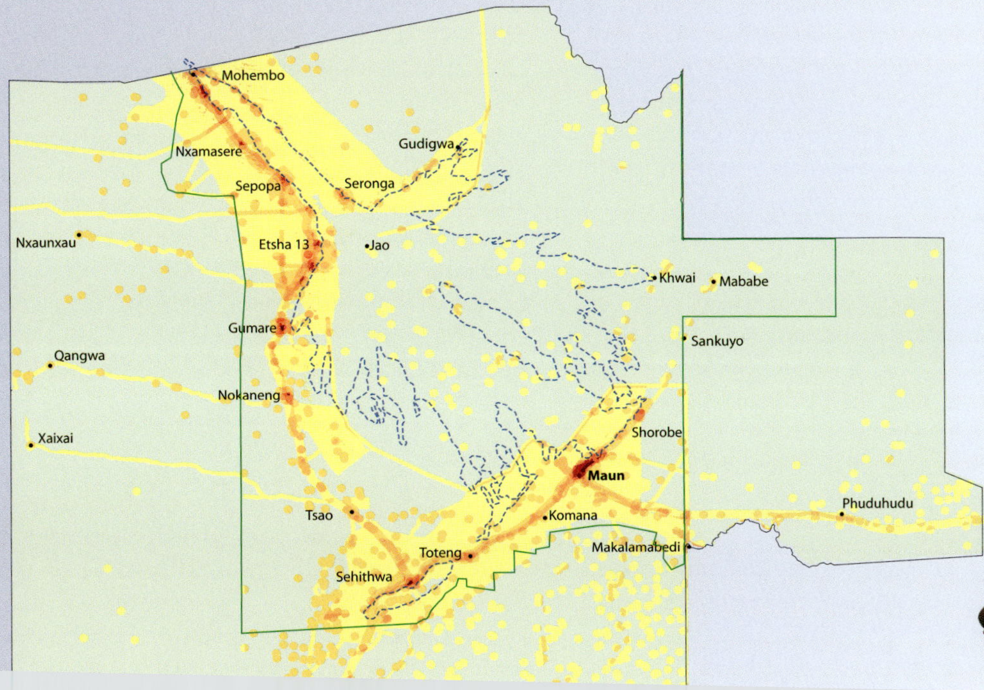

services reducing disease and mortality among children. Prior to the establishment of clinics and hospitals the people of Ngamiland were often the victims of malaria, gastro-intestinal infections, sleeping sickness, tuberculosis, venereal diseases, bilharzia and other maladies.

Populations throughout the world have grown, not because more children were born, but because more have survived. In fact, birth rates in Botswana have dropped significantly in recent decades. In 1981, the average number of children born to a woman in Botswana was 6.6, but this had halved to 3.2 children in 2006.[9] The decline in birth rate can largely be attributed to women attaining progressively higher levels of education, and therefore spending more of their lives in employment than at being mothers.

While declining birth rates have caused the overall expansion of the population to slow in recent years, mortality and illness from HIV/AIDS has probably had a substantial impact on growth. Infection rates in Botswana have been amongst the highest in the world for a long time, and about one-quarter of people aged 15-49 in Ngamiland were infected in 2004.[10]

Although infection rates may not have dropped, the situation has since improved because increasing numbers of people have access to anti-retroviral drugs and therefore survive. Before the drugs became widely available, life expectancy dropped from an average of 65 years in 1990–1995 to just 40 years in 2000–2005. Botswana was the first African country to provide antiretroviral drugs to everyone in need.[11]

No one lives permanently in most areas of the inner Delta and the remote areas of western and northern Ngamiland. Of the settlements that do exist, most are very small **(Figure 37)**, with a high proportion of the population clustered in a handful of larger villages and in the district capital of Maun. From census data gathered in 2001, the largest villages, each with more than 1,000 residents, were (from largest to smallest): Gumare, Shakawe, Etsha 6, Etsha 13, Seronga, Nokaneng, Sepopa, Sehithwa, Nxamasere, Mohembo West, Tsao, Matlapana and Xakao. In total, these 13 larger villages together with the town of Maun were home to about 84,800 people, or 58 % of Ngamiland's population. Since 1981, Maun's rate of growth has exceeded 6% per year, with the population doubling every 11 years. Other settlements, such as Shakawe and Gumare, have also grown rapidly.[12]

As a result of migrations over a long period of time, a large variety of people labelled as belonging to different ethnic and language groups live in Ngamiland. The most populous groups are the Tawana (also called BaTawana, most of whom live in Maun and various large villages), Wayeyi (mostly in smaller settlements along the southern and western margins of the alluvial fan, Hambukushu (mainly in settlements and villages along the western Panhandle) and Herero (largely in the south-west of the alluvial fan in such places as Sehithwa and Tsau). Other, smaller groups are Tcanikhoe (also spelled Xanekhwe or //Anikhwe), Bugakhwe (alternatively Bukakhwe), Deti (or Teti), Hura (or Ura), Ju/'hoansi, Gomahing, BaKgalagadi, BaKhurutsi, Masubia, Makalaka, Dxeriku, and various Europeans and Asians.

Some of the rapidly changing circumstances of peoples' livelihoods are illustrated in the table on the next page, which compares the results of the 1991 census with those of 2001. For example, whereas the majority of homes (78%) were traditional structures with thatched roofs (*built with labour*) in 1991, almost half had corrugated iron roofs (*built with cash*) in 2001. Two out of five people were employed in agriculture or the harvesting of natural resources in 1991, compared to about one in six people in 2001. Over three quarters of households had access to piped water in 2001, compared to about half in 1991. Other measures that have changed a good deal over the last few decades, and continue to change, are improved levels of education (with more people going to school and/or completing higher levels of education), greater access to health care, and higher levels of possession of imported manufactured goods, such as vehicles, telephones, radios and other domestic appliances.

Farming

Two types of agriculture predominate in Ngamiland: livestock and crop farming, and both are largely practised using traditional methods which depend on family labour and local natural resources such as water, soil nutrients, pastures and browse.[13] Much of this farming is 'subsistence' in nature, providing local residents with food for domestic consumption and, in the case of cattle, capital security. Approaches are thus quite different from those used in commercial farming operations where most inputs, for example fertilisers,

	1991	2001
Homes with piped water or using communal taps	52%	77%
Households using river water	17%	9%
Homes with corrugated iron roofs	18%	46%
Homes with thatched roofs	78%	49%
People employed in public administration	7%	19%
People employed in wholesale and retail enterprises	9%	15%
People employed in farming and harvesting natural resources	40%	15%

The Okavango Ramsar site has about 194,000 cattle and 99,000 goats.[14]

Flood recession farming is mainly practised along the western edges of the alluvial fan between the Etsha villages and Tsao, and also in the south-east between Shorobe and the upper reaches of the Boteti River.

farm managers, livestock feed, pesticides, electricity and water for irrigation, are specially purchased or hired.

Livestock are generally based at cattle-posts or the homes of their owners (especially goats) from where they graze outwards on a daily basis. Cattle and goats are the most abundant stock, with smaller numbers of sheep and donkeys. Most stock are held on communal or tribal land, and the great majority of animals are kept south of the buffalo veterinary fence. Less than half of all rural families have cattle and goats. For example, 37% of households in Ngamiland owned cattle in 2001, while 43% owned goats. Livestock ownership is thus very unequal, and the biggest herds or flocks also belong to relatively few owners. Goats are slaughtered infrequently for household consumption or sold when their owners need additional cash. With the exception of Herero farmers who frequently market cattle for commercial gain,[15] most people keep cattle as security assets and for occasional ceremonial use, for example when a family member marries or dies.

Fields are divided into those planted with rain-fed or dryland crops, and those used for flood-recession or *molapo* cultivation. By the year 2000, some 48,900 hectares had been cleared for crops in Ngamiland, of which 75% were dryland fields and 25% were fields in floodplain areas.[16] However, only about 10,000 hectares are used in any given year, the remainder being abandoned or left fallow.

Each *molapo* field (plural *melapo*) averages about two hectares, and is cleared on ground that slopes down into channels or broader floodplains. Maize, sorghum and vegetables, such as beans and pumpkins, are planted most often, usually in late winter and spring as temperatures begin to warm. The crops are planted in strips parallel to water lying in the channels or floodplains to benefit from floodwater moisture remaining in the soil. Although early rains in October and November provide supplementary water, the success of *molapo* farming is primarily determined by flooding – both the previous season's flooding and that of the coming season. Either too much or too little flooding is detrimental to *molapo* crops.

Yields from *melapo* are generally higher than those from dry lands, which can only be planted after the first good rain fall. Dryland crops also frequently suffer from shortages of water as a result of both limited and episodic falls of rain, and the low capacity for water retention in soils that are mainly sandy. Average yields for dryland crops are 162 kilograms of maize, 121 kilograms of sorghum and 144 kilograms of millet per hectare. These yields amount to less than US$100 per hectare if they are translated into values that would be paid for packaged cereals.

Although their methods and commodities differ, livestock and crop farming share several features.

The first is that these activities largely supplement the livelihoods of most rural families who live off a range of different incomes. Farming is just one source of income, and it is often small compared to cash earnings, remittances and social benefits. For instance, it was estimated in 2003 that at least 50%, and perhaps as much as 76%, of all rural income in Ngamiland was not generated by farming activities. Another set of figures showed farming to be the most important livelihood activity for less than one in four rural households.[17] Agriculture is therefore not the mainstay livelihood for most people in Ngamiland, and its importance will continue to diminish as people increasingly seek cash-based incomes in towns.

Secondly, these farming strategies are typical of a 'low input, low output' system. For example, farmers seldom invest in, or take measures that many outside observers would assume as necessary for better production. Few farmers thus add fertilizers, compost or manure to improve soil fertility, and weeds are not removed as often as they might be. As a result, crop yields are low. So too, are off-take rates of livestock that are often allowed to graze freely, when herding might provide better forage for the animals.

The major reason for low inputs is due to a third feature of farming in Ngamiland. This is the significant risk of failure or loss, which means that extra investment or effort often doesn't pay off. For example, pests and diseases plague both crops and livestock. Rinderpest wiped out most cattle in 1896, and lung sickness (Contagious Bovine Pleuropneumonia [CBPP]) effectively did the same in 1995 and 1996.[18] Earlier on, over a quarter (28%) of all cattle starved to death during the dry years that lasted between 1982 and 1988. Outbreaks of tripanosomiasis and foot-and-mouth disease occur from time to time. Crops are attacked by various parasites, stripped by locusts and red-billed quelea birds, and suffer from shortages of rain or floodwater. For example, Botswana declared droughts in 27 of the 33 years between 1964 and 1997.[19] Wildlife exerts further tolls on farming, for example by damaging crops and killing livestock. Indeed, the increasing incidence of such conflicts may undermine goodwill towards wildlife and the conservation of the Delta.

A fourth similarity is that it is usually impossible for most farmers to earn reasonable amounts of money because of low levels of production, and because market opportunities for surpluses are limited. This was not a problem before cash became essential, but nowadays everyone needs cash for necessities such as decent clothing, medicines, cell phones and efficient transport. And it is this need for money that now drives so many people to forsake farming for urban livelihoods where they have reasonable chances of earning incomes from employment or enterprises.

Finally, both livestock and crop farming enjoy very substantial subsidies from government. For example, crop farmers can get support to obtain draught animal power, animal-drawn farm implements, fencing materials, water tanks, fertilisers, threshing machines, mini-silos, chaff cutters, scotch carts, and canoes and paddles. Grants are also available for stock farmers to obtain poultry and guinea fowl, sheep and goats, equipment for boreholes and wells, fodder, fodder barns, dip tanks, kraals, crushes and loading ramps.[20] In addition, extension and veterinary officers provide farmers with free advice, veterinary medication and soil testing services. Very poor families can receive food baskets.

Rural life in Ngamiland would be even tougher without all these subsidies. Traditional farming clearly provides some food and security, but does little to provide most residents with the necessities of modern life. However, it is also true that farming is valuable to poor families that have few or no other sources of income. For example, a recent assessment of poverty found that about 28% of all households were below the poverty datum line.[21] What proportions of these were in rural and urban areas is not known. A challenge for the future is to evaluate whether the best options for alleviating poverty really lie in farming or other economic livelihoods. Likewise, we need to consider whether farming (and what kind of agriculture) or other enterprises are apt to make the best use of land in and around the Delta.

Land uses

All land in Ngamiland is either state-owned or communal, which is often also called tribal land **(Figure 38)**. Government departments directly manage state land while the Tawana Land Board is responsible for the administration and allocation of communal land. The state land consists of the Moremi Game

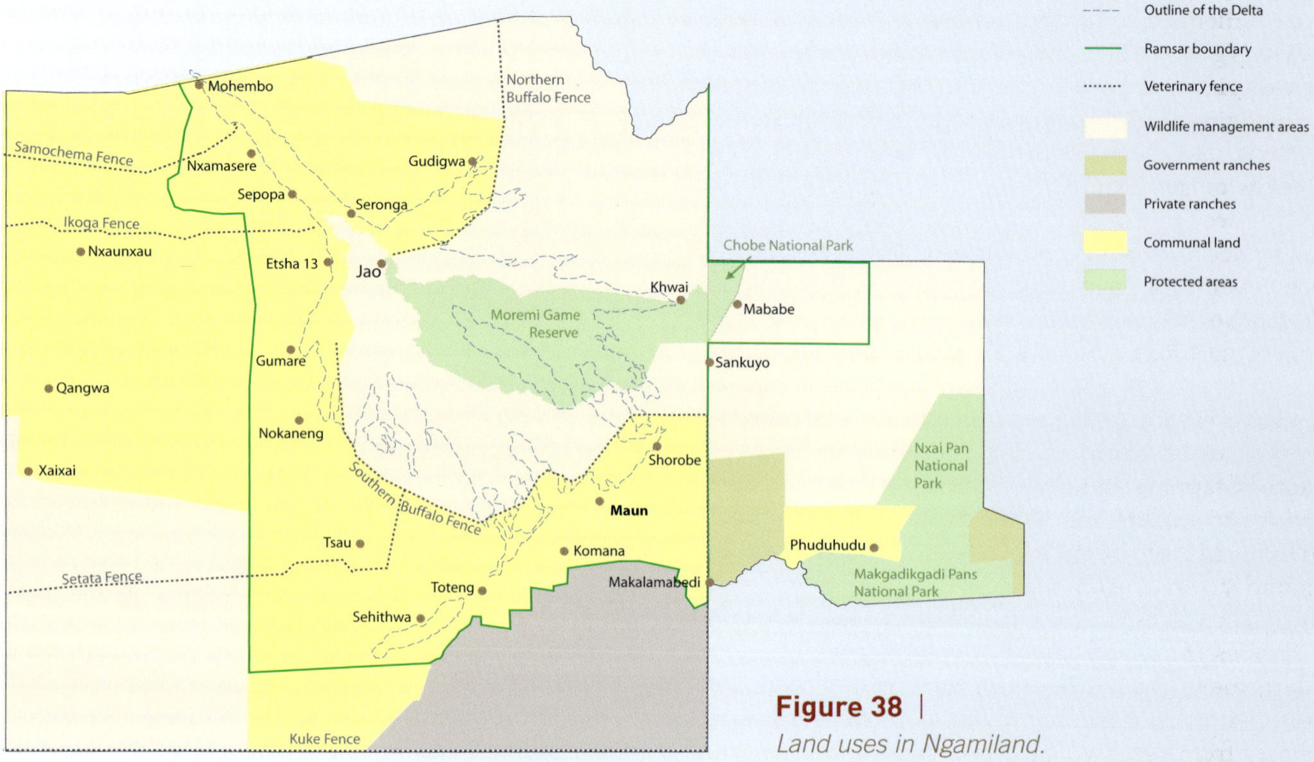

Figure 38 |
Land uses in Ngamiland.

Veterinary cordon fences were erected to limit the spread of disease between wildlife and livestock and to control infections between domestic animals when outbreaks occur. For instance, foot-and-mouth disease is easily transmitted from buffalo to cattle, and an epidemic of this disease could jeopardise the whole of Botswana's export beef industry.

However, veterinary (and other) fences are detrimental environmentally. This is particularly true in arid landscapes where animals need to roam over large areas to obtain water and forage because rainfall is erratic in space and time. As a result of the southern Buffalo fence, regular wildlife migrations from the Delta westwards towards Namibia and southwards to Lake Ngami have been stopped. Dramatic declines in numbers of wildlife in the Kalahari are probably due to their migrations being cut off, especially by the Kuke fence **(Figure 38)**. In addition, fences create biological islands where animals are prevented from inter-breeding with populations elsewhere. Genetic diversity is therefore reduced, while in-breeding may have unwelcome effects.

Reserve (covering 4,871 square kilometres), the 344 square kilometres of Chobe National Park that fall within Ngamiland, several large cattle ranches (used for experimental breeding, artificial insemination and quarantining) and the town of Maun.

In terms of the Tribal Grazing Land Policy (TGLP) of 1975, a large area of designated communal or tribal land was divided into commercial ranches. Most of the ranches each cover between 4,000 and 7,000 hectares. While the ranches were allocated as leaseholds to individual farmers, they have effectively become private property.

Land uses in the remaining communal areas are divided between those where emphasis is placed on crop and livestock farming and those where the primary use of land is for wildlife and tourism. The latter are called Wildlife Management Areas (WMAs), and each is known by a unique number, such as NG/21 or NG/33. Their boundaries and broad purposes were introduced in 1992 as part of a community-based natural resource management (CBNRM) programme. The goals of CBNRM are to provide local communities with benefits from wildlife and simultaneously to maintain large areas in and around the Delta for tourism and conservation. Many of the WMAs also provide buffers between farming areas and protected wildlife zones.

Of 29 WMAs in Ngamiland, rights for the use of 15 are currently leased to private entrepreneurs who use the areas for tourism and/or trophy hunting. Another 13 WMAs are allocated to communities, which usually enter into joint-venture agreements with tourism and/or hunting enterprises.[22] A variety of benefits accrue through the joint ventures. For example, local residents gain incomes from employment by the enterprises, which also pay royalties to community management trusts for tourism and hunting rights. Other benefits include support to local social services, such as schools and clinics, and the distribution to local residents of meat from hunted animals.

The economic conditions of people in some WMAs have improved substantially, especially in those areas where benefits are shared between relatively small numbers of households, and where the joint ventures have high commercial value for tourism and trophy hunting. For example, there were very few jobs or sources of cash in the settlements of Sankuyo, Khwai and Mababe prior to them establishing joint ventures, which now provide employment to about 50% of all the resident adults.[23] While the agreements with tourism and hunting operators may impose restrictions on land uses in certain areas, local residents are generally free to practice traditional farming. Cattle are, however, not allowed in the northern WMAs because of veterinary controls.

Communal land provides the poor with places where they can live and farm for free. But communal land is also free for the rich, which often leads to severe over-grazing. As the first president, Seretse Khama, observed in 1975: 'Under our communal grazing system, it is in no one individual's interests to limit the number of his animals. If one man takes his cattle off, someone else moves his cattle in'.

The boundaries of the Ramsar site run around the Delta and enclose a substantial area of Ngamiland. This is one of the largest Ramsar sites in the world, and was so designated in 1997 in recognition of its 'significant value for Botswana and… for humanity as a whole… because of its international significance in terms of ecology, botany, zoology, limnology or hydrology.'[24] The Ramsar site includes Botswana's frontage on the Linyanti Swamps and Kwando River. Since most of these wetlands lie across the border, it would be of value if Namibia were to extend the Ramsar site into its territory.

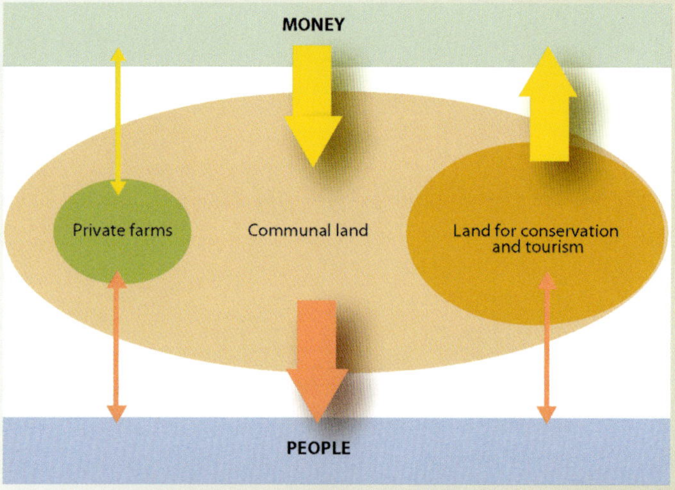

Figure 39 | *Major land use areas, and the flows of money and people into and out of these areas. All land in Ngamiland was originally communal for farming and the harvesting of natural resources. Recently, some land has been given to private commercial farms, while a larger area was set aside for the primary purpose of conservation and tourism and hunting enterprises. Large sums of money flow to sustain families in the remaining communal areas where it is hard to make a decent living, and many people therefore leave to live and work in urban areas. By contrast, large sums of money flow out of profitable wildlife and tourism areas, but critical questions are often asked about the beneficiaries of the revenues. Few questions are raised about the merits of private ranches, however. The narrow arrows reflect the small flows of people and money in and out of private farms, and of people in or out of tourism and wildlife areas.*

Figure 38 provides a perspective on the spatial extent of land uses. Another way of looking at this is provided by **(Figure 39)** which illustrates the flows of revenue and people associated with different land uses.

The allocation of large areas in and around the Delta for tourism and conservation by the Botswana government has been good for international tourists, the economy of the country, for investors in tourism, and for naturalists who value wildlife and wild places. However, it is often stated that the use of land for wildlife and tourism has been at the expense of rural livelihoods because residents have lost access to natural resources that they harvested traditionally.[25] Areas available for farming, collecting reeds, fishing and hunting, for example, have shrunk. In addition, the number of rural residents has grown, and so less land is available to support more people. Legal restrictions on the use of natural resources have also increasingly been introduced.

The status quo of land allocation in Ngamiland and its uses is frequently lamented. It is strongly argued that the loss of resources is unfair because conservation largely serves the interests of foreign tourists and of investors outside the Delta, many of whom are not Botswana citizens. Moreover, the great majority of beneficiaries are the white owners, senior employees and shareholders of tourism businesses.

Much of this is true, although increasing numbers of black citizens of Botswana are earning revenues from the same tourism industry, largely in the service sectors but increasingly as shareholders and employees in tourism enterprises. For example, a significant proportion of Maun's economy revolves around tourism with its knock-on effects, and the great majority of Maun's approximately 60,000 residents are Botswana citizens.

Furthermore, rural livelihoods are not lost to tourism and conservation as generally and significantly as critics often claim. There are three reasons for this. First, natural resources are lost to other land uses as well, for example through the allocation of communal land to private farms which are given free to select owners, many of whom are not from Ngamiland. Some communal land has also been set aside for government uses that are unrelated to tourism and conservation. And natural resources that should be for the sole use of rural residents, at least from an ethical point of view, are often used disproportionately by relatively wealthy people from towns. They frequently invest some of their savings in cattle which they place at rural cattle posts where their large herds consume much of the available

The Delta attracts about 100,000 visitors each year, largely to about 60 camps and lodges. Revenues earned from them, combined with those from trophy hunters, make up the bulk of Botswana's GDP now generated by tourism. Tourism is thus a major sector in the country's economy, coming second only to diamonds in terms of revenue generated. About 40% of the employment available in Ngamiland is provided by the tourism industry.[26] Some jobs directly serve visitors at lodges and camps or as poling guides, for example, while others indirectly support the tourism industry through services, such as those in retail outlets and the public service. Botswana's tourism policy has three principle goals: (a) to encourage tourists who occupy permanent accommodation, as opposed to casual campers; (b) to maximise financial returns from tourism for the people of Botswana, especially those who tolerate the costs of living close to wildlife, and (c) to ensure that tourism is carried out in an ecologically pasture. Local, poorer residents with small herds are placed at a significant disadvantage. Frequent bush fires (*see page 84*) and the on-going clearing of land for crops also contribute to the loss of natural resources.

Second, the so-called *dependence* of most rural families on local natural resources is simply exaggerated. As described earlier (*see page 118*), many rural residents gain most of their income as cash from sources that are independent of local natural resources that they may use. Ironically, much of the off-farm income is also a product of tourism and conservation enterprises, the very activities that supposedly limit the economic health of local households. Thus, many salaries, royalties, remittances, social benefits and government subsidies are funded by taxes derived from enterprises that make money from tourists and wildlife.

Third, arguments that rural people should have more land for farming and other forms of natural resource harvesting assume (a) that decent livelihoods can be made from traditional land uses and (b) that rural life is preferable. Both suppositions are tenuous. Even where land is available, farming is difficult for the reasons discussed above (*see page 119*). Fish yields are also very low (*see page 103*) and there are few

People living in rural areas are much more diverse that is commonly perceived. Some are members of rich and large families, others are small and poor. The former usually depend on off-farm incomes for their daily needs, while poorer people depend much on farming and the use of other local resources. In achieving a balance between conservation, the economic value of tourism and trophy hunting, and uplifting the livelihoods of local residents of the Delta, it seems important that these poorer families be given the best opportunities of using natural resources. There is also a need for benefits derived from wildlife to be more evenly spread across the Delta's population.

markets where occasional surpluses can be sold. And this remains true despite significant government subsidies to support small-scale farming and fishing. Rural life also appeals only to certain people, whereas the great majority of schooled men and women are attracted to urban lifestyles.

Debates on the pros and cons of tourism/conservation vis-à-vis the necessity of maintaining rural livelihoods are not easy to conclude. The contentious issues may also be viewed from different angles and framed differently, for example by asking if local interests should prevail over wider ones. Are long term goals more important than those that meet immediate needs? More directly: 'Whose Delta is it?' and 'How long should the Delta's resources remain useful?' And are traditional livelihoods preferable to modern ones and, if so, preferable to whom?

The Botswana government recently grappled with these tough questions while compiling the Okavango Delta Management Plan, which was completed in 2008, and the Ngamiland Integrated Land Use Development Plan, finalised in 2009.[27] Both plans emphasise the economic and conservation values of the Delta's natural resources for the country, international community and local residents, and the need to maintain the natural processes that keep the Delta functioning, as described in Chapter 4. Proposals are made to enhance and distribute the benefits of these resources more fairly. The plans also recommend the expansion of zones for wildlife management, and to physically separate land uses where the potential for conflict is high.[28] These include conflicts between farming and wildlife, and between different kinds of farming. The challenge now remains for these and other useful recommendations to be implemented.

KEY POINTS

1. The population of Ngamiland has grown rapidly as a result of medical services that have improved survival, but growth rates have recently slowed because fewer children are born nowadays.
2. Crop and livestock farming produce low yields, and are risky as a result of disease, pests, shortages of rain, and too much or too little flooding. Inputs to farming are therefore low.
3. In addition to low agricultural production, rural livelihoods are tough as a result of limited access to services, cash incomes and modern necessities.
4. Whilst farming is valuable to very poor families, most income in rural areas is not from farming or from the direct harvesting of natural resources, but rather comes from salaries, business earnings, remittances, social benefits and subsidies.
5. Rates of urbanisation have been high as increasing numbers of people, especially younger men and women, have moved to seek salaries and services in towns. Over half the population now lives in urbanized settings in towns and large villages.
6. The daily activities of most people therefore now focus on earning money, on purchasing food (rather than harvesting it), on buying clothes (rather than making them), on quick, comfortable transportation (rather than walking) and on fast electronic communication (as opposed to the 'bush telegraph').
7. Likewise, modern medicine has largely replaced traditional cures, and food security has taken the place of food self-sufficiency. Formal education is held in high esteem, and its teachings are more highly regarded than folklore. Likewise, public services and government play stronger roles than traditional leadership.
8. Most land is used for three purposes: communal farming, private ranching, and wildlife conservation and tourism.
9. There is considerable debate over the beneficiaries and comparative value of communal land and that used for tourism and conservation.

8

Floods for the future

The Delta has been protected from the kinds of development that often lead to the loss of natural resources. What can be done to secure its good health in the foreseeable future?

Okavango Delta: Floods of Life had its origins in a programme led by the International Union for the Conservation of Nature.[1] A major goal of this IUCN programme was to promote the use of information to monitor the abundance and composition of life in the Delta. What are known as *freshwater biodiversity monitoring* programmes have been implemented elsewhere in the world, particularly along rivers where the deterioration of water quality upstream is likely to cause populations of plants and animals to decline downstream. Such changes are used to flag the possibility that upstream polluters, for example, have added additional or new poisons to their effluent. That information is then passed to people responsible for managing the quality of river water so that the sources of pollution can be tracked down and eliminated.

Many projects to do similar monitoring in the Delta have been implemented. As a first step, baseline information was collected to determine what populations of animals and plants were present. In such a diverse environment, samples had to be taken in different places, and samples had to be collected at various times to account for seasonal changes. Quantitative indices as measures of both the abundance and diversity of organisms were then developed so that future samples could be compared against the baseline indices. Changes in the indices should thus provide warnings that all is not well, either as a result of water being altered locally or as it flows out of Angola and Namibia.

This book was intended to synthesise the baseline monitoring information and experience, and thus to demonstrate the value of monitoring freshwater biodiversity in the Delta. However, it was found that more work was needed to develop appropriate monitoring measures before a synthesis could be attempted.[2] An alternative approach was then taken, which was to summarise what is known about the Delta, in a format that would interest people who should care for the Delta. In other words, since an adequate *supply* of information on freshwater biodiversity monitoring was not yet available, we hoped that this book would create greater *demand* for monitoring and responsive management in the future. Likewise, we hope that the thoughts offered here will raise awareness about the challenges to be faced.

Threats: internal and external

These first few paragraphs touched on the intentions of IUCN and other people to keep an eye on the Delta. For what, however, should the monitoring be? Thoughts on that question are split in two categories: internal risks that originate within and immediately around the wetland, and external factors that potentially affect the Okavango River in its Angolan catchment and its passage through Namibia to Mohembo. We begin with potential internal hazards.

Chemical pollution

Thus far this has not been a major problem in the Delta, perhaps because contamination has not been significant and because wetlands filter out organic compounds so effectively. However, there are concerns about effluent from villages and towns, lodges, camps and boats for tourists. Measures have recently been implemented to require tourism facilities to process sewage water and to bury the remaining concentrates beneath saline sediments in the centres of islands (*see page 56*). Some lodge and camp owners have also implemented their own monitoring programmes to check the quality of water around their facilities.

The use of pesticides to kill tsetse flies has caused considerable alarm. There have been two major recent applications of deltamethrin, in 2001 and 2002, when the entire alluvial fan was sprayed several times from aeroplanes. Deltamethrin was used because it targets invertebrates and degrades rapidly, unlike such persistent and infamous poisons as DDT and dieldrin that further accumulate in deadly concentrations in animals at the top of the food web. The use of deltamethrin is also an improvement over the use of endosulphan (which is toxic to fish) during the 1980s for tsetse fly control in the Delta.[3]

After the deltamethrin spraying, the abundance of aquatic and terrestrial invertebrates was found to have dropped by between 25 and 65%, depending on the family of invertebrate family. Dragonflies (*see page 95*) and beetles were particularly hard hit, while others were less affected.[4] By 2003, many groups had recovered to pre-spraying abundances; and while some species had not recovered, others were more abundant than before the spraying. It is not known if this was due to natural population changes or if the 'new' species occupied niches left vacant by those that were killed. Fish and birds were apparently not harmed by the spraying.

Although contamination by pesticides applied within Botswana has perhaps not been too problematic, this potential hazard needs to be monitored continuously since the impacts of inadvertent poisoning can be utterly devastating, and because residues can remain toxic for so long. Contaminants applied locally could also increase toxic burdens introduced from upstream and atmospheric sources of chemical poisoning.

Alien species

Invasive, exotic plants and animals are another potential threat to the Delta. Like chemical pollution, they may be introduced unwittingly and remain inconspicuous, at least in the beginning. And as with pollutants, there are some problems already.

Foremost of these is the water weed *Salvinia molesta*, which first appeared in the Delta in the mid 1980s. Experiments by Namibians along the Kwando River showed that the weed could largely be contained by *Cyrtobagous salviniae*, a small beetle introduced from Brazil. The Botswana Department of Water Affairs accordingly implemented an on-going programme to

The most serious invasive alien is the water weed Salvinia, mats of which can cover large areas of open water. The weed was first recorded in the Delta in July 1986 at Xini Lake. Adult beetles (above) destroy leaf buds of Salvinia and their larvae tunnel into the rhizomes causing the plants to disintegrate and sink.

introduce the insect, as well as using manual labour to remove mats of Salvinia. So far, these efforts have limited the spread of the weed in the Delta, but the control measures have to be applied repeatedly wherever Salvinia reappears. A good deal of vigilance is thus needed to detect outbreaks and to respond to them in good time. One difficulty arises from the growth of beetle populations apparently lagging behind that of Salvinia, which allows mats of the weed to develop. Fortunately, the low nutrient content of Delta water (*see page 53*) appears to limit the growth and spread of the weed.

A variety of other alien plants occur in the Delta, such as thorn apples or jimson weeds (*Datura ferox* and *D. stramonium*), the burweed (*Xanthium stramonium*), *Mimosa pigra*, *Sesbania* species, syringa (*Melia azederach*) and exotic food plants, such as guavas, pawpaws and cassava. Thorn apples and burweed sometimes cover large areas of disturbed ground in the Delta (*see page 82*), and all these species may be invasive at the expense of natural vegetation.

Unlike most large wetlands elsewhere in the world, no invasive species of fish yet occur in the Delta.[5] However, there are risks of the unwanted introduction of such species as the Nile tilapia (*Oreochromis nilotica*), which might hybridise with several of the tilapia species that are native to the Delta. The possibility of Nile tilapia reaching the Delta from the Zambezi River (through the Chobe River, Linyanti Swamps and Selinda Spillway) appears slight, but the route down the Okavango River is direct and easy to follow should people in Angola or Namibia start to farm with these fish.

Clearing of channels

Development in most societies strives towards stability and predictability, which in the case of the Okavango Delta is in direct conflict with its inherently unstable, shifting nature. Channelling of water and clearing waterways has been done frequently in the past, and is still often proposed to facilitate boat traffic and to provide villages with permanent water supply. However, the removal of blockages will make the channels more permanent which, in turn, will create stable flooding patterns. This is likely to reduce the size of the Delta in the long term, as well as lowering the productivity of floodplains for livestock, wildlife, fish and other populations. Bush encroachment may increase, and biological diversity could be reduced.

Moreover, and perhaps fortunately, clearing activities are usually doomed to fail because they attempt to reverse natural changes. The failure of every historically recorded channel clearing effort over the last century in the Delta bears this out.

Elephants

Populations of these giants have been booming in northern Botswana for the past 20 years. Large swathes of mature riverine forest along the Kwando and Okavango Rivers (in Namibia's Bwabwata National Park) have been entirely decimated, and elephants have done the same to mopane woodlands in some areas of the Delta (*see page 87*). If similar significant damage happens to woodlands on islands there is the risk of transpiration and the concentration of salts being reduced (*see page 55*), and that could lead to the Delta becoming saline.

Although almost everyone condemns the loss of woodland and agrees that there are too many elephants, no one appears willing to take measures to control the problem. Any significant culling programme would elicit howls of protest from animal rights groups, and this could damage Botswana's reputation and tourism industry.[6]

External challenges in the catchment and elsewhere

Large areas of land and natural resources in the Delta have essentially been set aside for conservation. This is economically valuable to the country as a whole and financially beneficial to many local residents. At the same time, this approach does much to safeguard environmental health.

Quite different perspectives and values hold in the Okavango River catchment in Angola and Namibia. Here, little land is allocated for environmental conservation, few people benefit financially from tourism, and more important economic interests elsewhere in the two countries carry much greater weight (*see page 35*).

It is possible that we may one day understand how to manage water flows and flooding patterns to maintain production, but this will require major improvements in our understanding of nutrient cycling and other aspects of the Delta's physiology. Until then, centres of production need to carry on wandering across the floodplains as they do now, unhindered and unmanaged.

Most conservation programmes focus on rare, charismatic or iconic species. But plants (*Chapter 5*) and animals (*Chapter 6*) should look after themselves as long as the functioning (*Chapter 4*) of this wetland is not altered. It is the system functioning of the Delta that is so special, and what we all marvel at is simply the manifestation a big, healthy ecosystem working in an unimpaired way.

The paradox of the Delta is that its productive value lies not in the water, but in the nutrients that the same river system, animals and aerosols have deposited here over at least the last 120,000 years (*see page 24*). Yet, the water and nutrients cannot be separated. Water is needed to release the productive potential of nutrients, but without the nutrients the water would be rather lifeless.

In recent years, several studies and projects have explored development options in the catchment area in Angola and for the flow of Okavango River water as it passes through Namibia.[7] With few exceptions, the postulated development scenarios have focused on (a) options for water use, (b) government policies and programmes for the use of water and (c) water allocations for the three countries. These are all extremely important aspects.

Yet little or no attention has been paid to (a) the financial benefits of tourism and wildlife in Angola and Namibia, (b) the growing use (and likely abuse) of the catchment's resources by private people and companies, (c) the promotion of the Okavango River Basin as a valuable ecological entity, and (d) the sharing of benefits between Botswana, Namibia and Angola. These, too, are important matters, to which we later return.

Namibia

Official perspectives on the Okavango in Namibia concentrate on the use of river water for agricultural, domestic and industrial use and fish farming. Although several non-government organisation (NGO) programmes have promoted non-consumptive uses of the Okavango in Namibia and entrepreneurs have established a number of lodges and camps, the value of the river in attracting tourism revenue is seldom mentioned in government circles.

For the future, we can expect more water to be extracted from the river, both by government projects and private use. Usage will increase gradually for much of the time, but it will also escalate sharply when Namibia next experiences conditions as dry as those in the early 1990s. Plans to pipe limited volumes of water to the central areas of the country were then developed, but were suddenly shelved after good rains fell.

The plans also attracted harsh criticism from many environmental groups who were concerned about the impacts of reduced flow into the Delta. Similar criticism was voiced when Namibia investigated the construction of a hydro-power facility at Popa Falls (*see page 21*).

An important point, however, is that the Namibian population, economy, and demands for water have all grown considerably since the 1990s. Thus, when very

Angola

Much of the Angolan catchment remains the *terras do fim do mundo*; the land at the end of the earth, unknown and unseen by most members of the public. The servants of the public in government will probably place emphasis on the development of water resources and more fertile soils elsewhere, which are closer to the more populous regions of the country. Potential for hydro-power development has been identified at several sites along tributaries of the Okavango, but serious plans to develop these have yet to be made. It is possible that 'run-of-the-river' designs could be used to minimize the impacts of hydropower schemes on water flow.

Since rainfall is usually higher and less variable, and evaporation lower in Angola than in Namibia, abnormally dry conditions in the catchment are unlikely to be that demanding of river resources. However, as in Namibia and Botswana, the relatively small population in the catchment will continue to grow steadily, expanding its footprint on the catchment. And as in the other countries, much of the footprint will be from low input, low output traditional farming.

While common sense dictates that the relatively infertile soils and remoteness from markets should make most of the Angolan catchment unsuited to commercial crop production, entrepreneurs may see the broad floodplains of rivers as suitable for large-scale irrigation, perhaps for sugar cane, maize, rice or jatropha, for example. Such developments will doubtless have to be subsidised with public money, thus copying the unprofitable irrigation schemes in Namibia (*see page 35*). Foreign investments and influences may be involved, as is happening elsewhere in several African countries, where large tracts of land are leased to feed foreign consumers and to meet growing demands for biofuels.

The amounts of water used for each irrigation scheme should be relatively small. However, the cumulative offtake of several farms – together with those in Namibia[8] – could be significant, especially if large volumes are removed at the start of the growing season in September and October when river levels are lowest (*see page 41*).

Of greater concern is what might run off all these irrigation schemes. Since the soils are low in nutrients, substantial applications of fertilizers will be required will be needed, and if applied on a large scale, significant residues will find their way into the river water and Delta. A recent study of residues of DDT, aldrin and HCB (hexachlorobenzene) found significant concentrations in sediments in the Delta.

The available evidence indicates that these poisons were washed down the Okavango River into the Delta. Interestingly, progressively higher concentrations were found downstream, so the lowest residues were in the Panhandle and the highest were in Lake Ngami and other distal parts of the alluvial fan.[9]

Erosion off the large irrigation schemes, together with the clearing of land for small-scale cropping could lead to increased loads of suspended clay in the river water. These are unlikely to have an immediate impact on the Delta since much of the clay would be filtered out before reaching the sandy, permeable substrate of the alluvial fan. Over the longer term, however, increasing clay inputs would reduce the chances of the Delta remaining a freshwater system (*see page 53*).

Finally, global forces are likely to shape the Delta in ways that are both unforeseen and less manageable. Climate change is an obvious example. The bedfellows of economics and politics will influence food and energy prices (and therefore motives for irrigation schemes) and the fickle tourism industry, for example, by raising safety concerns and eroding savings for holidays. And tectonic movements inside the Delta may shift the distribution of water in ways that surprise its residents.

Challenges for management

A host of factors can thus compromise the health of the Delta. Some are global in scale, others are founded hundreds of kilometres upstream in the Okavango River catchment. Yet others are local in origin. How should people with responsibilities for managing the Delta respond, and what should be done by those with direct economic and financial interests in the Delta? Likewise, what can the international community do to maintain the wealth and health of this apparent organism (*see page 39*)?

Keeping the overall system dynamic

The biological wealth of the Delta, along with the goods, services and capital it provides is very largely a product of water supply being able to vary across two dimensions: time and space. The pulsing of inflows between seasons and years results in parts of the Delta being inundated and then dried. Similarly, the spatial redistribution of water allows some areas to dry up and others to be flooded afresh. These changes continually shift the interface between water and nutrients, enabling the former to mobilise the latter. The changes also allow desiccation, grazing, fire and decomposition to recycle nutrients into forms available for new bouts of biological production.

Little of this would happen if the changes in time and space were to stop. If the water remains in one place, nutrients there will be locked up in plants, while dormant nutrients elsewhere will lie quietly, their potential locked away by the absence of water. A key measure for the future is therefore to allow the interplay between water and accumulated nutrients to continue moving.

Building demand for information and response

The chapter began by describing how information to monitor the Delta was being developed. The technical challenges in doing this for such a dynamic wetland are enormous. A further challenge lies in the use of monitoring information. Without special measures to develop procedures for its use (including accountability), all that we can *hope* for is that: managers will get the information, will understand its importance, and will respond accordingly.

The same is true for planning. For instance, the recently completed Okavango Delta Management Plan (ODMP) has been hailed as a good example of how planning processes should be done. It also makes excellent recommendations, which everyone *hopes* will be implemented.[10]

But hoping is not enough. Neither is it enough to have lots of information and plans in our shop window if there are no customers. There are no simple methods of developing demand for monitoring (or any other objective) information and implementation. A multitude of approaches are required, many of which will have to be repeated and tuned to different audiences. Accountability and incentives are perhaps the most important facets, as are public awareness, lobbying and political interests.

Enlarging the identity and security of the Delta

In combination, no other wetland in Africa is (a) as large, (b) well protected, (c) so pristine, (d) of such high economic value and (e) as well-known internationally. These attributes probably set the Delta apart from all other African wetlands by an order of magnitude. Indeed, the combination of values could be unique worldwide.

The Okavango Delta is thus a big deal, and it deserves a reputation larger than the one it now enjoys. It also requires an identity that goes beyond its present borders. Its ownership should surpass national pride since this is an international asset, which has been called the Jewel of the Kalahari (which covers a large area of Africa) and the Eden of Africa (which is a significant part of the world).[11]

We close by exploring three aspects that are crucial to a better, more secure future for this very significant wetland.

Going beyond food

Generalised assumptions that agriculture is Africa's salvation are unfortunate, particularly when they are applied to low input, low output farming in areas that are poorly suited to food production and its commercialization. Along with these broad assumptions go a variety of notions, for example that rural Africa is best suited to traditional economies, that nothing better can be done with communal land, that secure tenure is unnecessary or too complicated in communal areas, and that alternative livelihoods are just too remote and challenging. As a result and for example, development programmes vigorously promote food self-sufficiency. Yet virtually every young person in Africa wants food security and cash security, and thus makes the one-way journey from a rural to an urban environment.[12]

Much of the debate about the pros and cons of financial and economic benefits of conservation and tourism are bedeviled by assumptions that low input, low output,

security for the many poor families in Ngamiland. But surely the development and expansion of farming should be done in places where it is productive, where farmers and families can make a decent living from their harvests, and where the environmental costs of land clearing and grazing are worth bearing?

As discussed previously (*see page 124*), this debate is not easy to resolve. But we do suggest that the debate be tackled earnestly (and honestly, on the basis of economic and environmental arguments) when the use of the Delta's natural resources are considered. Exactly the same objective approach is needed when people ponder over land uses in the Angolan catchment and along the Okavango River in Namibia.

Broadening OKACOM

OKACOM is the river basin commission established between Angola, Namibia and Botswana in September 1994.[13] During its first 10 years, the commission concentrated largely on the concept of water allocations between the countries, and on over-seeing various donor-funded projects. As senior public servants, mostly in departments of water based in Luanda, Windhoek and Gaborone, its commissioners had limited opportunity to engage in day-to-day matters of concern to the Okavango River and its Delta.

More recently, OKACOM has done much more to strengthen and broaden its mandate in the complicated playing field that may see sovereign interest, state responsibility and good neighbourliness as rival themes.[14] Some of these developments were stimulated by Botswana's initiative in having the Delta declared as a Ramsar site, and in formulating the Okavango Delta Management Plan (ODMP).

Other progress has been made through a range of projects conducted under the auspices of OKACOM. These include work to develop (1) support among local communities for the environmental health of the River Basin; (2) approaches to integrated water resource management; (3) trans-boundary diagnostic assessments (TDAs); (4) a strategic action programme (SAP); (5) studies of environmental flows (E-flows). In addition, an OKACOM secretariat has been established in Maun, which is alongside a newly created office for the Botswana Directorate of Environmental Affairs. Such broadening of scope from one purely based on water supply to one in which water is seen as an ecological

In our view, OKACOM could further enhance its role by obtaining greater representation of local and environmental interests. A broadening of its debating field would also be valuable, especially if greater emphasis were placed on such issues as pesticide contamination, the enrichment of river water by nutrients from farming, private uses of the Okavango's resources, and the creeping expansion of the human footprint on the Okavango River Basin. It would be useful for OKACOM to ensure that environmental impact assessments precede all major developments in the Basin. Regular environmental audits commissioned by OKACOM would provide useful measures to assess the use of water and all other natural resources. The same audits should take stock of benefits of these resources, reporting on which are useful, which require promotion, and how the benefits are used. To reiterate, the human footprint around the Delta and the source of water on which it depends is expanding. The expansion and its implications must be monitored, and OKACOM is the only organisation with the mandate to do the monitoring. By taking on these kinds of responsibilities, OKACOM could evolve from its present advisory capacity into one where it becomes a river basin authority.

In essence, discussions about the interests of Angola, Namibia and Botswana have largely focused on what the three governments do, relationships between them, and on the sharing of water. There has been less discussion on what local people and private enterprises in the River Basin might think. Likewise, focus has been placed on the value of water, rather than on the environmental value of the Basin. For the future, expanded perspectives that reach across national boundaries to recognize the unitary value and integrity of the Okavango River Basin are desirable. In the longer term, each of the three countries stands to gain a great deal from such trans-national perspectives.[16]

Seeing beyond Mohembo

Much of the debate within Botswana about the management of the Delta concentrates on issues in and around the Delta. This is rightly so. But the debates sometimes include the implication that Botswana alone controls the destiny of the Delta. That is misguided, however. For example, the economy of the Delta depends largely on the willingness of foreigners to pay for tourism, and how well the Delta can compete with other foreign holiday attractions. More importantly, the quality and quantity of water inflow is controlled entirely by Namibia and Angola. The Delta is thus surrounded by external influences, with which Botswana should engage.

Broadening the identity of the Delta is necessary for such engagement. Most people now see the Okavango and Delta as the same thing or place, as if only the 'Okavango Delta' matters. The word *Okavango* has virtually been unwittingly commandeered for exclusive use in Botswana, as if the Delta is independent of its Angolan catchment. But the broader area of the Okavango River Basin is actually what counts, both in its own right as a basin covering 192,500 square kilometres, and as the Delta's source of water. And unintended implications that the Delta equals the Okavango diminish value of its catchment. But there is a grand opportunity here, which is to capitalize on the international fame of the Delta and to spread that reputation (and its associated values) across the whole river basin. In essence, use the 'big deal' name of the Delta to give the Okavango River Basin 'big deal' status. Doing so would give greater security to the Delta's water sources by reducing the possibility of the catchment being put to degrading use.

Spreading the fame and brand name is one thing. Another is to extend the vested interests and experience that now lie in the Delta upstream into the Angola. Tourism companies from Botswana, in conjunction with the Botswana government, could give value and protection to the catchment by investing in tourism and wildlife in Angola. Jobs and infrastructure would be created there, and new perspectives on the use of water and natural landscapes would be developed in Angola. Food security for residents of the catchment could come from wildlife, tourism and its associated industries, rather than from slash-and-burn farming that produces low yields.

Having the Botswana government as a partner would give security and credibility to the private sector. Angola's reputation would be enhanced by giving the land at the end of the earth renown and

Much has been learnt about the Delta over the years. Further study will expand our knowledge so that humanity will be better placed to care for the health of this wetland and the value of its wealth.

purpose. Botswana would be the first African country to invest in the protection of a foreign environmental resource. Benevolent organizations and people elsewhere in the world who believe in the value of the Okavango could contribute to these developments. In doing so, they would swell the international community of people who would be vigilant over the entire Okavango.

Elements of these recommendations are to be found in the proposed Kavango Zambezi Trans-Frontier Conservation Area (KAZA).[17] This initiative involves Angola, Botswana, Namibia, Zambia and Zimbabwe, and aims to broaden and connect conservation areas over some 300,000 square kilometres. Tourism is to be promoted, in part to boost the livelihoods of residents of KAZA. While the KAZA area encompasses the Delta, only part of the Cuito River and its tributaries is included in Angola.

Implementation of the ideas suggested above would also have the private and public sectors of one country invest in another country, and they would focus on development of the Okavango River Basin as a hydrological and ecological entity. Amongst other measures, the current Ramsar site could be extended across the whole Basin.

The whole Okavango River Basin should also be developed into a tourism destination. In addition to visiting individual attractions in the catchment, tourists could also follow the flow of water by boat, road and air from Tchicala Tcholohanga to Menongue, Calai, Andara, Jao and Lake Ngami, for example. How many other places in the world offer journeys and holidays across an entire river basin, much of which is pristine, wild and scenic? The Delta would of course gain by having much greater security over the quantity and quality of water on which it depends. The national economy of Botswana would have greater confidence in counting on future revenues from tourism and conservation.

It is our hope that vision and innovation prevails to expand the concept of the Delta and the resources that are central to its health. This will be to the benefit of the diatoms (*see page 82*), cladocerans (*page 100*), people (*page 109*) and all others who use the Delta's nutrients and waters. It can be done!

KEY POINTS

1. While the Delta is in relatively pristine condition, it faces local threats from chemical contamination, invasive aliens, the clearing of channels, and a burgeoning population of elephants.
2. Since the Delta depends on water from the Okavango's Angolan catchment that passes through Namibia, it may suffer from increasing water extraction for irrigation, and the addition of pesticides, nutrients and clay to water that enters Botswana.
4. Global forces, such as climate change, food and energy prices, and demands for tourism will directly and indirectly influence the future of the Delta.
5. Uncontrolled pulses and shifts in the supply of water across the Delta are crucial for the maintenance of its biological wealth and health.
6. Measures to improve the use of information and the implementation of plans are required to enhance the management of the Delta.
7. Expanding the identity of the Okavango upstream of the Delta would add security to its future, as well as enhancing the value of natural resources and livelihoods in the catchment.
8. Trans-boundary discussions on management of the Okavango should be broadened to include, for example, local interests, environmental impact assessments and audits, and the sharing of benefits from the Okavango.
9. Investments in the catchment by the Botswana tourism industry and government would help protect the Delta's future.
10. Bold, innovative management of the Delta and its catchment will allow those who come after us to benefit from this eden and jewel of the world.

Notes and sources

Chapter 1: Introducing the Okavango Delta

1. Ross K. 1999. *Okavango: jewel of the Kalahari*. Struik, Cape Town.
2. VanderPost C. 2005. Early maps of Ngamiland and the Okavango Delta. *Botswana Notes and Records* 37: 196–207.
3. Adapted from definitions proposed by the American Geological Institute, http://www.agiweb.org.
4. The Ramsar List was established in response to Article 2.1 of the Convention on Wetlands (which was signed and agreed to at Ramsar, Iran in 1971).

Chapter 2: Shaping the Delta

1. Adapted from McCarthy TS. 1992. Physical and biological processes controlling the Okavango Delta – a review of recent research. *Botswana Notes and Records*. 24: 57–86.
2. Based on a map compiled by the Council for GeoSciences, South Africa and provided by Spike McCarthy, and data provided by USGS National Earthquake Information Center, http://neic.usgs.gov/neis/epic/epic_rect.html.
3. From NASA Landsat and Modis images.
4. Adapted from Burrough SL, Thomas DSG, Shaw PA & Bailey RM. 2007. Multiphase Quaternary highstands at Lake Ngami, Kalahari, northern Botswana. *Palaeogeography, Palaeoclimatology, Palaeoecology* 253: 280–299; and Burrough SL, Thomas DSG & Bailey RM. 2009. Mega-Lake in the Kalahari: A Late Pleistocene record of the Palaeolake Makgadikgadi system. *Quaternary Science Reviews* 28: 1392–1411. The satellite image was taken on 1 September 2009 and provided by NASA/GSFC, MODIS Rapid Response (http://rapidfire.sci.gsfc.nasa.gov).
5. Substantial layers of material formed from the decomposition of the bedrock have been found in north-western Ngamiland, suggesting that alluvial and aeolian sedimentation may be less significant than usually assumed (see McFarlane MJ, Coetzee SH, Kuhn JR, Vanderpost CHM & Eckardt FD. 2007. In situ rounding of quartz grains within an African surface weathering profile in North West Ngamiland, Botswana. *Zeitschrift für Geomorphologie* 51: 269–286.)
6. Moore AE, & Larkin PA. 2001. Drainage evolution in south-central Africa since the breakup of Gondwana. *South African Journal of Geology* 104: 47–68.
7. Ringrose S, Vanderpost C & Matheson W. Geomorphic Origins of the Okavango Delta (alluvial fan) and adjacent dune deposits in northern Botswana. Unpublished manuscript.
8. Same as 7.

Chapter 3: The Okavango River: flow of a lifeline

1. Based on an interpolation of average seasonal totals calculated from rainfall records obtained from the Global Historical Climate Network database, and the Botswana and Namibia Meteorological Services.
2. Liebenberg PJ. 2009. *Technical report on irrigation development in the Namibia section of the Okavango River Basin*. Ministry of Agriculture, Water & Forestry, Windhoek, Namibia
3. Based on the total estimated population in the Basin area of each country as a percentage of the national population. The map is based on Morebodi BBH. 2001. *Botswana National Atlas*. Department of Surveys and Mapping, Gaborone; Mendelsohn JM, Jarvis AM, Roberts CS & Robertson T. 2002. *Atlas of Namibia*. David Philip, Cape Town; and the UNEP database of population density in Angola (grid2.cd.usgs.gov/globalpop/Africa).
4. Ramberg L & Wolski P. 2008. Growing islands and sinking solutes: processes maintaining the endorheic Okavango Delta as a freshwater system. *Plant Ecology* 196: 215–231.
5. *Diagnóstico Transfronteiriço Bacia do Okavango – Análise Socioeconómica Angola*. Report for Transboundary Diagnostic Assessment, EPSMO/OKACOM project; and Mendelsohn JM. 2006. *Farming systems in Namibia*. RAISON, Windhoek.
6. Swatuk LA. 2005. Whose values matter most? Water and resource governance in the Okavango River Basin. *Mechanisms of Economic Regulation* 35: 521–529.
7. Mendelsohn JM. 2009. *Land use in Kavango: past, present and future*. Report for Transboundary Diagnostic Assessment, EPSMO/OKACOM project.

Chapter 4: The functioning Delta

1. Gumbricht T & McCarthy TS. 2002. Hierarchical processes and patterns sustaining the Okavango: an integrated perspective for policy and management. In Bernard T, Mosepele K & Ramberg L, (eds). *Environmental Monitoring of Tropical and Subtropical Wetlands*. Harry Oppenheimer Okavango Research Center. Okavango Report Series No. 1.
2. Much of the information on water flow was provided by Piotr Wolski, HOORC.
3. Calculated from data collected by the Department of Water Affairs at Mohembo.
4. Wolski P, Savenije HHG, Murray-Hudson M & Gumbricht T. 2006. Modelling of the flooding in the Okavango Delta, Botswana, using a hybrid reservoir-GIS model. *Journal of Hydrology*. 331: 58–72.
5. Gumbricht T, Wolski P, Frost P & McCarthy TS. 2004. Forecasting the spatial extent of the annual flood in the Okavango delta, Botswana. *Journal of Hydrology* 290: 178–191.
6. McCarthy TS, Ellery WN & Stanistreet LG. 1993. Lakes of the north eastern region of the Okavango Swamps, Botswana. *Zeitschrift für Geomorphologie* 37: 273–294.
7. MODIS images provided by NASA/GSFC, MODIS Rapid Response (http://rapidfire.sci.gsfc.nasa.gov).
8. McCarthy TS, Cooper GRJ, Tyson PD, Ellery WN. 2000. Seasonal flooding in the Okavango Delta, Botswana – recent history and

future prospects. *South African Journal of Science* 96: 25–33; Mazvimavi D & P Wolski. 2006. Long-term variations of annual flows of the Okavango and Zambezi Rivers. *Physics and Chemistry of the Earth*, 31: 15–16, 944–951; Wolski P, Gumbricht T & TS McCarthy. 2002. Assessing future change in the Okavango Delta: the use of a regression model of the maximum annual flood in a Monte Carlo simulation. In Bernard T, Mosepele K & Ramberg L, (eds). *Environmental Monitoring of Tropical and Subtropical Wetlands*. Harry Oppenheimer Okavango Research Center. Okavango Report Series No. 1; Tyson PD, Cooper GRJ & McCarthy TS. 2002. Millennial to multi-decadal variability in the climate of southern Africa. *International Journal of Climatology* 22: 1105–1117.

9. Same as 5 above.
10. From a two-way analysis of variance by Piotr Wolski. Another 25% of variance is not explained and interaction between the three variables accounts for a further 3%.
11. From analyses of MODIS images by Piotr Wolski, HOORC.
12. Wolski P, Savenije HHG, Murray-Hudson M & Gumbricht T. 2006. Modelling of the flooding in the Okavango Delta, Botswana, using a hybrid reservoir-GIS model. *Journal of Hydrology* 331: 58–72.
13. From data analysed by Piotr Wolski, who used a long series of LandSat images to determine the frequency of inundation. The procedure for identifying water is described in Wolski P & Murray-Hudson MH. 2006. *Reconstruction of 1989–2005 inundation history in the Okavango Delta, Bostwana from archival Landsat imagery*, Globwetland Symposium, ESA-ESRIN, Frascati, Italy, 19–20 October 2006 (available from http://www.globwetland.org/news_links/links/433855kj.pdf).
14. Wolski P & Gumbricht T. 2003. *Mapping hydrological units in the Okavango Delta*. Report for Water and Ecosystem Resources for Regional Development (WERRD) project. Maun, Botswana.
15. Wilson BH. 1974. Some natural and man-made changes in the channels of the Okavango Delta. *Botswana Notes and Records* 5, 132–153; Shaw PA. 1983. Fluctuations in the level of Lake Ngami: the historical evidence. *Botswana Notes and Records* 5: 132–84.
16. Wolski P & Murray-Hudson M. 2006. Recent changes in flooding in the Xudum distributary of the Okavango Delta and Lake Ngami, Botswana. *South African Journal of Science* 102: 173–176.
17. This was adapted from a Quickbird image used by Google Earth, while the extent of flooding in other years was mapped from MODIS images provided by NASA/GSFC, MODIS Rapid Response (http://rapidfire.sci.gsfc.nasa.gov).
18. Ellery WN, Ellery K, Rogers KH, McCarthy TS & Walker BH. 1993. Vegetation, hydrology and sedimentation processes as determinants of channel form and dynamics in the northeastern Okavango Delta, Botswana. *African Journal of Ecology* 31: 10–25.
19. Wolski P & M Murray-Hudson. 2008. An investigation of permanent and transient changes in flood distribution and outflows in the Okavango Delta, Botswana. *Physics and Chemistry of the Earth* 33: 157–164.
20. Redrawn from data assembled by Piotr Wolski, HOORC.
21. McCarthy TS, Stanistreet LG & Cairncross B. 1991. The sedimentary dynamics of active fluvial channels on the Okavango fan, Botswana. *Sedimentology* 38: 471–487.
22. Ramberg L & Wolski P. 2008. Growing islands and sinking solutes: processes maintaining the endorheic Okavango Delta as a freshwater system. *Plant Ecology* 196: 215–231; Masamba WRL, Huntsman-Mapila P, Ringrose S & Wolski P. 2008. Hydrological, geomorphological and geochemical Processes in the Okavango Delta: water resource management implications. *Proceedings of the Second IASTED Africa Conference September 8–10, 2008*
23. Ramberg L, Wolski P & Krah M. 2006. Water balance and infiltration in a seasonal floodplain in the Okavango Delta, Botswana. *Wetlands*, 26: 3, 677-690.
24. Ramberg L & Wolski P. 2008. Growing islands and sinking solutes: processes maintaining the endorheic Okavango Delta as a freshwater system. *Plant Ecology* 196: 215–231.
25. Gumbricht T, McCarthy J & McCarthy TS. 2004. Channels, wetlands and islands in the Okavango Delta, Botswana, and their relation to hydrological and sedimentological processes. *Earth Surf Proc Land* 29:15–29.
26. Adapted from Ramberg L & Wolski P. 2008. Growing islands and sinking solutes: processes maintaining the endorheic Okavango Delta as a freshwater system. *Plant Ecology* 196: 215–231
27. Bauer-Gottwein P, Langer T, Prommer H, Wolski P & Kinzelbach W. 2007. Okavango Delta Islands: interaction between density-driven flow and geochemical reactions under evapo-concentration. *Journal of Hydrology* 335: 389–405; and Zimmermann S, Bauer P, Held R, Kinzelbach W & JH Walther. 2006. Salt transport on islands in the Okavango Delta: Numerical investigations. *Advances in Water Resources* 29: 11–29.
28. Bonyongo MC. 2004. *The ecology of large herbivores in the Okavango Delta, Botswana*. PhD Thesis, Bristol University, Bristol, UK.
29. Cronberg G, Gieske A, Martins E, Prince Nengu J & Stenström L-M. 1996. Major ion chemistry, plankton, and bacterial assemblages of the Jao/Boro River, Okavango Delta, Botswana: the swamps and flood plains. *Archiv für Hydrobiologie Supplement* 107: 335–407.
30. Adapted from Krah M, McCarthy TS, Huntsman-Mapila P, Wolski P & Sethebe K. 2006. Nutrient budget in the seasonal wetland of the Okavango Delta, Botswana. *Wetlands Ecology and Management* 14: 253–267.
31. Garstang M, Ellery WN, McCarthy TS, Scholes MC, Scholes RJ, Swap RJ & Tyson PD. 1998. The contribution of aerosol- and water-borne nutrients to the functioning of the Okavango Delta ecosystem, Botswana. *South African Journal of Science* 94: 223–229.
32. Krah M, McCarthy TS, Huntsman-Mapila P, Wolski P & Sethebe K. 2006. Nutrient budget in the seasonal wetland of the Okavango Delta, Botswana. *Wetlands Ecology and Management* 14: 253–267.
33. Mubyana T, Krah M, Totoloz O & Bonyongo MC. 2003. Influence of seasonal flooding on soil total nitrogen, organic phosphorus and microbial populations in the Okavango Delta, Botswana *Journal of Arid Environments* 54: 359–369
34. Same as 29.
35. Hogberg P, Lindholm M, Ramberg L & Hessen DO. 2002. Aquatic food web dynamics on a floodplain in the Okavango Delta, Botswana. *Hydrobiologia* 470: 23–30. Lindholm M, Hessen DO, Mosepele K & Wolski P. 2007. Food webs and energy fluxes on a seasonal floodplain: The influence of flood size. *Wetlands* 27: 775–784.
36. Hancock P, Oake K, Tyler SJ & Brewster C. 2005. *Monitoring waterbirds at Lake Ngami – an Important Bird Area: June 2004 to January 2007*. Reports to Wetlands International and Wilderness Wildlife Trust.

Chapter 5: Plants: producers, filters and distributors

1. From data provided by Mike Murray-Hudson HOORC.
2. A great deal of the information available on the Delta's plants was compiled by the pioneering work of Pete Smith and Fred and Karen Ellery.
3. Ramberg L, Hancock P, Lindholm M, Meyer T, Ringrose S, Sliva J, Van As J & Vanderpost C. 2006. Species diversity of the Okavango Delta, Botswana. *Aquatic Sciences* 68: 310–337.
4. Ellery WN & Tacheba B. 2003. Floristic Diversity of the Okavango Delta, Botswana. In Alonso, L.E. and L-A. Nordin (eds.) A Rapid Biological Assessment of the Aquatic Ecosystems of the Okavango

5. Junk W, Brown M, Campbell I, Finlayson M, Gopal B, Ramberg L & Warner B. 2006. The comparative biodiversity of seven globally important wetlands: a synthesis. *Aquatic Sciences* 68: 400–414.
6. Adapted from Meyer T. 1999. *Ecological mappings in the research area of the Harry Oppenheimer Okavango Research Centre, Okavango Delta, Botswana.* MSc thesis, Hochschule Anhalt, Bernburg, Germany.
7. From Smith P. 1989. In Snowy Mountains Engineering Corporation (SMEC) Cooma NSW Australia. *Ecological Zoning Okavango Delta. Final report, Volume 1.* Main report to the Ministry of Local Government and Lands, Government of Botswana.
8. Sliva J, Murray-Hudson M & Heinl M. 2004. Vegetation species diversity inventory in the Okavango Delta, Botswana. In Mosepele K. & Mosepele B (eds.) *AquaRap II: A Rapid Assessment of the Aquatic Ecosystems of the Okavango Delta, Botswana: Low Water Survey.* Draft Final Report, Conservation International, Washington DC, USA, pp. 69–83.
9. The orchid *Habenaria pasmithii* may be the Delta's only endemic plant.
10. Same as above.
11. For the first time in years, waters of the Okavango and Kwando met on the Selinda on the 18th of August 2009 at 18.52052 South, 23.45602 East (Alison and Roger Heath, personal communication).
12. Smith P. 1989. In Snowy Mountains Engineering Corporation (SMEC) Cooma NSW Australia. *Ecological Zoning Okavango Delta. Final report, Volume 1.* Main report to the Ministry of Local Government and Lands, Government of Botswana; Ellery WN & Tacheba B. 2003. Floristic Diversity of the Okavango Delta, Botswana. In Alonso, LE & Nordin L-A (eds.) A Rapid Biological Assessment of the Aquatic Ecosystems of the Okavango Delta, Botswana: High Water Survey. *RAP-Bulletin of Biological Assessment* 25: 69–96; Sliva J, Murray-Hudson M & Heinl M. 2004. Vegetation species diversity inventory in the Okavango Delta, Botswana. In: Mosepele K & Mosepele B (eds.) AquaRap II: *A Rapid Assessment of the Aquatic Ecosystems of the Okavango Delta, Botswana: Low Water Survey. Draft Final Report.* Conservation International, Washington DC, USA, pp. 69–83; Murray-Hudson M. 2009. *Floodplain vegetation responses to flood regime in the seasonal Okavango Delta, Botswana.* Ph D thesis University of Florida.
13. Mendelsohn JM & el Obeid S. 2004. *Okavango River: the flow of a lifeline.* Struik, Cape Town. Based on NDVI data collected over 16 seasons between 1985/1986 and 2002/2003.
14. Iron is reduced to Fe^{2+}; nitrogen changes from nitrate (NO_3^{2-}) to NH_3 (via many intermediates) and may be lost to the atmosphere, as also happens after the reduction of sulphur to H_2S and carbon to CH_4 (methane).
15. Same as above.
16. As a slight diversionary hypothesis, the north-eastern area of mopane woodlands may have experienced greater clay sedimentation from waters of the Zambezi River when it flowed into this area of Botswana (see page 24). By contrast, inflows to the western areas may have carried less clay because they largely came from areas of Kalahari Sand.
17. Burrough SL, Thomas DSG, Shaw PA & Bailey RM. 2007. Multiphase Quaternary highstands at Lake Ngami, Kalahari, northern Botswana. *Palaeogeography, Palaeoclimatology, Palaeoecology* 253: 280–299; Burrough SL, Thomas DSG. 2008. Late Quaternary lake-level fluctuations in the Mababe Depression: middle Kalahari palaeolakes and the role of Zambezi inflows. *Quaternary Research* 69: 388–403.
18. Cholnoky BJ. 1966. Die Diatomeen im Unterlaufe des Okawango-Flusses. *Beihefte Nova Hedwigia* 21: 1–102.
19. Mackay AW, Todd M & Wolski P. 2009. *Monitoring and simulating threats to aquatic biodiversity in the Okavango Delta.* Project report for the Darwin Initiative, University College of London.
20. Heinl M, Neuenschwander A, Sliva J & VanderPost C. 2006. Interactions between fire and flooding in a southern African floodplain system (Okavango Delta, Botswana). *Landscape Ecology* 21: 699–709; and Heinl M, Sliva J, Murray-Hudson M & Tacheba B. 2007. Post-fire succession on savanna habitats in the Okavango Delta wetland Botswana. *Journal of Tropical Ecology* 23: 705–713.
21. Cassidy L. 2003. *Anthropogenic Burning in the Okavango Panhandle of Botswana: Livelihoods and Spatial Dimensions.* MSc Thesis, University of Florida, Gainsville, FL, U.S.A; and Heinl M. 2005. Fire Regime and Vegetation Response in the Okavango Delta, Botswana. PhD thesis, *Lehrstuhl fur Vegetaionsökologie, Department für Ökologie, Technische Universität München*, Munich, Germany.
22. Tacheba B, Segosebe E, VanderPost C & Sebego R. 2009. Assessing the impacts of fire on the vegetation resources that are available to the local communities of the seasonal wetlands of the Okavango, Botswana, in the context of different land uses and key government policies. *African Journal of Ecology* 47: 71–77.
23. Frost PGH, Plavsic M & Rogow T. 2004. Observations on the occurrence, impacts and effects of fire in the Okavango Delta In: Mosepele K, & Mosepele B. (eds.) *AquaRap II: A Rapid Assessment of the Aquatic Ecosystems of the Okavango Delta, Botswana: Low Water Survey. Draft Final Report.* Conservation International, Washington DC, USA, pp. 98–111.
24. Ramberg L, Lindholm M, Bonyongo C, Hessen DO, Heinl M, Masamba W, Murray-Hudson M, VanderPost C & Wolski P. In press. Aquatic ecosystem responses to fire and flood size in the Okavango Delta – Natural experiments on seasonal floodplains. *Wetland Ecology and Management.*

Chapter 6: Animals: shifting consumers

1. The species are allocated to their preferred habitats but, of course, some species occur in more than one habitat. Adapted from Ramberg L, Hancock P, Lindholm M, Meyer T, Ringrose S, Sliva J, Van As J & VanderPost C. 2006. Species diversity of the Okavango Delta, Botswana. *Aquatic Sciences* 68: 310–337.
2. The title of Frans Lanting's book: *Okavango: Africa's Last Eden*, published in 1993 by Chronicle Books, San Francisco.
3. Darwall WRT, Smith KG, Tweddle D & Skelton P (eds) (2009). *The Status and Distribution of Freshwater Biodiversity in Southern Africa.* Gland, Switzerland: IUCN and Grahamstown, South Africa: SAIAB.
4. Much of this information was supplied by Jens Kipping. His research follows that of the pioneering work of Eliot Pinhey.
5. Ramberg L, Hancock P, Lindholm M, Meyer T, Ringrose S, Sliva J, Van As J. & VanderPost C. 2006. Species diversity of the Okavango Delta, Botswana. *Aquatic Sciences* 68: 310–337.
6. Okavango Crocodile Research Group. 2008. *A Management Plan for the Conservation of the Nile crocodile (*Crocodylus niloticus*) in the Okavango Delta, Botswana.* University of Stellenbosch, South Africa.
7. Pete Hancock (personal communication) and Hancock P, Mpofu Z, Tyler SJ & Meyer T. 2005. *A baseline survey of the slaty egret in the Okavango Delta Ramsar site, Botswana.* Report by BirdLife Botswana for Okavango Delta Management Plan.
8. Hancock P, Oake K, Tyler SJ & Brewster C. 2005. *Monitoring waterbirds at Lake Ngami – an Important Bird Area: June 2004 to January 2007.* Reports to Wetlands International and Wilderness Wildlife Trust.
9. Lindholm M & Hessen DO. 2007. Competition and niche partitioning in a floodplain ecosystem: a cladoceran community squeezed between fish and invertebrate predation. *African Zoology* 42: 158–164; and Lindholm M, Hessen DO & L Ramberg. 2009. Diversity, dispersal and disturbance: Cladoceran species composition in the Okavango Delta. *African Zoology* 44:24–35.

10. Mosepele K, Moyle PB, Merron GS, Purkey DR & Mosepele B. 2009. Fish, Floods, and Ecosystem Engineers: Aquatic Conservation in the Okavango Delta, Botswana. *BioScience* 59: 53–64.
11. Welcomme, R.L., 2001. *Inland fisheries; ecology and management*. Blackwell Science, Oxford.
12. Mosepele K, Mmopelwa TG, Mosepele B. 2003. Characterization and monitoring of the Okavango Delta artisanal fishery. In Bernard T, Mosepele K & Ramberg L. (eds). Environmental *Monitoring of Tropical and Subtropical Wetlands*. Harry Oppenheimer Okavango Research Center. Okavango Report Series No. 1: 391–413.
13. Mosepele K & Kolding J. 2003. Fish stock assessment in the Okavango Delta: Preliminary results from a length-based analysis. In Bernard T, Mosepele K & Ramberg L (eds). *Environmental Monitoring of Tropical and Subtropical Wetlands*. Harry Oppenheimer Okavango Research Center. Okavango Report Series No. 1: 363–390; and Bayley PB. 1988. Accounting for effort when comparing tropical fisheries in lakes, river-floodplains, and lagoons. *Limnology and Oceanography* 33: 963–972.
14. Mosepele K. 2003. *Trends in Fisheries Development and Fish Utilization in the Okavango Delta*. Report for Water and Ecosystem Resources for Regional Development (WERRD) project. Maun, Botswana.
15. Caspar Bonyongo (personal communication) and Bonyongo MC. 2004. *The ecology of large herbivores in the Okavango Delta, Botswana*. University of Bristol, Bristol, UK.

Chapter 7: People: predators and protectors

1. VanderPost C. 2004. Human Sprawl and the African Wilderness of the Okavango. *South African Geographical Journal* 86: 65–73
2. From information in Lane P, Reid A & Segobye A. (eds). 1998. *Ditswa Mmung: The archaeology of Botswana*. Pula Press and The Botswana Society, Gaborone.
3. Tlou T. 1985. *A history of Ngamiland, 1750 to 1906. The formation of an African state*. Gaborone: MacMillan; Bock J. 1998. Economic development and cultural change among the Okavango Delta peoples of Botswana. *Botswana Notes and Records* 30: 27–44
4. Tlou T & Campbell AC. 1984. *History of Botswana*. Macmillan Botswana.
5. Kgathi DL, Ngwenya BN & Wilk J. 2007. Shocks and rural livelihoods in the Okavango Delta, Botswana. *Development Southern Africa*. 24: 289–308
6. IUCN. 1992. *The IUCN Review of the Southern Okavango Integrated Water Development Project*, Final Report, Gaborone
7. Recent figures come from the Central Statistics Office, Botswana. The last full census was in 2001 when a total of 124,700 people were counted in Ngamiland. The 2006 Inter-censal sample survey showed that the population to be 138,600. At an annual growth rate of 2.1%, Ngamiland's population should amount to 156,000 in 2010 and 185,000 in 2020.
8. VanderPost C. 2007. Protected areas in Ngamiland, Botswana: investigating options for conservation-development through human footprint mapping. *International Journal of Environmental Studies* 64: 555–570.
9. Results supplied by the Central Statistics Office based on the 1991 population census and 2006 inter-censal survey.
10. Central Statistics Office. 2004. *Botswana AIDS Impact Survey II: Preliminary Results*. Ministry of Finance and Development Planning, Gaborone.
11. http://www.avert.org/aids-botswana.htm.
12. Officially, these are designated as villages but they are really small towns.
13. integrated perspective for policy and management. In Bernard T, Mosepele K & Ramberg L (eds). Environmental *Monitoring of Tropical and Subtropical Wetlands*. Harry Oppenheimer Okavango Research Center. Okavango Report Series No. 1:
14. Department of Environmental Affairs. 2008. *Okavango Delta Management Plan*. Dept. of Environmental Affairs, Gaborone, Botswana.
15. Pastoralists rely on livestock for both their revenue streams and capital savings or security, whereas people with mixed economies (such as crops and livestock, or wages and livestock) usually keep their animals as security and use other income sources to meet day-to-day revenue needs. This is particularly true in societies living on communal land where people are prohibited from having capital assets and security in land.
16. Same as 13 above.
17. Kgathi DL, Bendsen H, Blaikie PEMJ, Ngwenya B & Wilk J. 2003. *Rural Livelihoods, Indigenous Knowledge Systems and the Political Economy of Access to Natural Resources in the Okavango Delta*. Water and Ecosystem Resources for Regional Development (WERRD) project; Kgathi DL, Ngwenya BN & Wilk J. 2007. Shocks and rural livelihoods in the Okavango Delta, Botswana. *Development Southern Africa* 24: 289–308
18. The outbreak of CBPP was so threatening to the whole beef industry that the Botswana government culled all cattle. An estimated 320,000 cattle were thus killed in Ngamiland.
19. Ministry of Finance and Development Planning. 1997. *Study of poverty and poverty alleviation in Botswana*. Gaborone.
20. Anon. 2006. *Agricultural Support Schemes*. Ministry of Agriculture, Gaborone.
21. Central Statistics Office. 2008. *Poverty Datum Line for Botswana 2003*. Government of Botswana, Gaborone.
22. Of the 15 WMAs leased to commercial operators, 11 are mainly for hunting and four are for tourism. Most of the 13 WMAs allocated to communities have private sector partners, eight being used largely for hunting and five for tourism. One WMA has not been designated.
23. Mbaiwa JE. 2009. Tourism development, rural livelihoods and biodiversity conservation in the Okavango Delta, Botswana. In Hottola P. (ed). *Tourism strategies and local responses in southern Africa*. CAB International 2009.
24. The Ramsar List was established in response to Article 2.1 of the Convention on Wetlands (which was signed and agreed at Ramsar, Iran in 1971).
25. Magole L & Magole LI. 2009. The Okavango: Whose Delta is it? *Physics and Chemistry of the Earth* 34: 874–880.
26. Turpie J, Barnes J, Arntzen J , Nherera B, Lange G-M & Buzwani B. 2006. *Economic value of the Okavango Delta, Botswana, and implications for management*. Okavango Delta Management Plan. Dept. of Environmental Affairs, Gaborone, Botswana.
27. Tawana Land Board and Northwest District Land Use Planning Unit. 2008. *Okavango Delta Ramsar site land use and land management plan (2005–2029)*. Published by Okavango Delta Management Plan Project Secretariat, Department of Environmental Affairs; and Department of Lands, Ministry of Lands and Housing. 2009. *Ngamiland Integrated Land Use Plan*. Gaborone, Botswana.
28. Darkoh MBK & Mbaiwa JE. 2009. Land-use and resource conflicts in the Okavango Delta, Botswana. *African Journal of Ecology* 47: 161–165.

Chapter 8: Floods for the future

1. Darwall WRT, Smith KG, Tweddle D & Skelton P. (eds). 2009.

2. Biodiversity monitoring studies in the Delta are reviewed by Dallas HF. 2009. *Wetland monitoring using aquatic macroinvertebrates.* Technical report for Biokavango Project, Harry Oppenheimer Okavango Research Centre, University of Botswana. The Freshwater Consulting Group, University of Cape Town, Cape Town, South Africa. The main difficulty with biodiversity monitoring stems from the dynamic condition of the Delta, particularly in its alluvial fan As a result, we do not expect, nor desire indices of biodiversity to be constant. The interpretation of any changes would also be hard since they could reflect changing natural processes or man-made degradation.
3. Games IP. 1981. *Report on the effects of a deltamethrin and endosulfan mixture on the non-target arthropods during the 1981 tsetse spraying program.* Maun, Botswana Department of Veterinary Services and Tsetse Fly Control; Fox, P.J. & Matthiessen, P. 1982. Acute toxicity to fish of low-dose aerosol applications of endosulfan to control tsetse fly in the Okavango Delta, Botswana. *Environmental Pollution Series A* 27: 129–142.
4. Perkins JS & Ramberg L (eds). 2004. *Environmental monitoring of Tsetse fly sprayingimpacts of the Okavango Delta – 2002.* Okavango Report Series, No. 2. Final Report prepared by the Harry Oppenheimer Okavango Research Centre, University of Botswana.
5. Junk W, Brown M, Campbell I, Finlayson M, Gopal B, Ramberg L & Warner B. 2006. The comparative biodiversity of seven globally important wetlands: a synthesis. *Aquatic Sciences* 68: 400–414.
6. One option that has been considered is the transfer of elephants to south-eastern Angola, where almost all were wiped out during the civil war. Significant numbers of elephants have indeed recently moved on their own into south-eastern Angola (Mike Chase personal communication).
7. The most comprehensive of these are the trans-boundary diagnostic assessments and environmental flow studies recently commissioned by the Environmental Protection and Sustainable Management of the Okavango River Basin (EPSMO) of the Global Enviornmental Facility (GEF) project.
8. Liebenberg PJ. 2009. *Technical report on irrigation development in the Namibia section of the Okavango River Basin.* Ministry of Agriculture, Water & Forestry, Windhoek, Namibia.
9. Mmualefe LC, Torto N, Huntsman-Mapila & Mbongwe B. 2008. Supercritical fluid extraction of pesticides in sediment from the Okavango Delta, Botswana, and determination by gas chromatography with electron capture detection (GC-ECD) and mass spectrometry (GC-MS). *Water SA* 34: 405–41.
10. Wolski P, Ramberg L, Magole L & Mazvimavi D. 2009. Evolution of river basin management in the Okavango system, Southern Africa. In Ferrier R & Jenkins A (eds). *Handbook of Catchment Management.* Wiley-Blackwell.
11. Ross K. 1999. *Okavango: jewel of the Kalahari.* Struik, Cape Town; and Lanting F. 1993. *Okavango: Africa's Last Eden.* Chronicle Books, San Francisco.
12. Food self-sufficiency aims for people to produce enough food to feed themselves, while food security implies that people can feed themselves adequately, irrespective of whether they grow or buy their food.
13. http://www.okacom.org
14. Ashton PJ, & Neal M. 2003. An overview of key strategic issues in the Okavango Basin. In Turton AR, Ashton PJ & Cloete TE. (eds). *Transboundary rivers, sovereignty and development: hydropolitical drivers in the Okavango River basin.* AWIRU, Pretoria, South Africa, Green Cross International, Geneva, Switzerland. Other papers in this volume are also relevant to these issues.
15. Same as 10 above.
16. Swatuk LA. .2005. Whose values matter most? Water and resource governance in the Okavango River Basin. *Mechanisms of Economic Regulation* 35: 521–529.
17. http://www.kazapark.com.

Photographic credits

Astronaut photography, Earth Sciences and Image Analysis Laboratory, NASA Johnson Space Center (http://eol.jsc.nasa.gov): Cover, 85 (middle left); **Biokavango project**: 102 bottom; **Tapson Wilfred Bombo**: 40; **Sven Bourquin**: 93 4th left; **Helge Denker**: 30 both on left & top three on right, 33, 34 top, 36; **Digital Globe**: 50; **Earth touch/www.earth-touch.com**: 42 top left, 97, Back flap, Back cover middle left; **Pete Hancock**: 3 bottom, 99 bottom middle, 102 top right, 119; **Alison & Roger Heath**: 74 top left and right, 77 all, 88 middle far left; **Mike Holding**: 2, 3 top, 16 top right, 42 right, 54, 57 both on right, 59, 63 bottom left, 67, 72, 85 bottom left, 87, 88 bottom right, 93 top left & right, 95 bottom left & far right; top right, 106, 117, 123, 131 bottom, 137, 138, Back cover middle centre; **Jens Kipping**: 19, 95 first 3 on the left & bottom middle; **Frans Lanting/www.lanting.com**: 1, 6 & 7, 8, 10, 16 right & bottom left, 17, 35, 38, 63 bottom right, 79 bottom, 105, 108, 110, 132, 140; **Tim & June Liversedge**: 18, 47 both, 61 right, 90, 99 bottom left, 102, 115 top, Back cover middle right; **Anson Mackay**, 83 all; **Wilma Matheson**: 60 bottom; **Graham Matthews**: 100 all; **John Mendelsohn**: 21, 26, 28, 30 bottom left, 31 all, 34 bottom, 37, 39, 42 bottom left, 52 both on left & 3rd on right, 57 top two on left & bottom left, 61 left, 63 middle left & top right, 64, 70, 80, 81 top, 82 both, 84, 85 top left & right, 88 top right, 109, 111, 114 all, 117, 121, 126, 129, 131 top right, Back cover top left; **Mark Muller & Ali Flatt**: 49, 52 2nd right, 60 top, 73 both, 74 bottom left, 85 bottom right, 86, 88 top left, 91 right, 98 bottom left, 99 bottom right, 101 bottom, 111, 112, 115, 118, 124; **Frances Murray-Hudson**: 88 3rd right; **Tor Naesje**: 7; **NASA Landsat and MODIS**: 23 all, 24, 44; **Richard Randall**: 5 top, 9, 14, 15, 52 bottom right, 53, 55, 57 3rd left, 58, 76, 78, 81 bottom, 88 bottom left & 2nd right, 93 2nd, 3rd & bottom left; both on bottom right, 98 top & bottom right, 99 both top & all 3 middle row, 104, 115 bottom, 116, 125, 129 bottom, 131 top left; **Gabor Salamon**: 63 top left; **Ben van der Waal**: 101 top; **Wilderness Safaris**: 5 bottom, 66, 74 bottom right, 75; **Wilderness Safaris (Dana Allen)**: 4, 79 top, 107, 127, Back cover bottom left, Front flap; **Wilderness Safaris (Dave Hamman)**: Back cover top right; **Piotr Wolski**: 52 top right.